TEACHING and LEARNING in AGRICULTURE

A guide for agricultural educators

Prepared by
Abraham Blum

Under the guidance and sponsorship of the
Extension, Education and Communication Service (SDRE)
Research, Extension and Training Division (SDR)

FOOD AND AGRICULTURE ORGANIZATION OF THE UNITED NATIONS
Rome, 1996

The designations employed and the presentation of material in this publication do not imply the expression of any opinion whatsoever on the part of the Food and Agriculture Organization of the United Nations concerning the legal status of any country, territory, city or area or of its authorities, or concerning the delimitation of its frontiers or boundaries.

M-67
ISBN 92-5-103750-7

Preface

It is generally acknowledged that agricultural education and training are of vital importance in promoting sustainable agricultural production and rural development. The role of the agricultural teacher, therefore, cannot be overemphasized. He or she occupies a unique position in the rural environment, primarily because of the positive impact that education and training can have on human resource development.

Teachers of agriculture are usually well-educated subject matter specialists who have in-depth knowledge of specific areas of science and technology. However, often because of their restricted preparation for teaching, they run into difficulties with the programme of instruction for which they are responsible. A sound technical knowledge of a specific field is essential but, to render it useful and relevant when applied to teaching and learning, it must be accompanied by an understanding as well as the practical application of pedagogical skills.

Unfortunately, many teachers of agriculture are lacking in their knowledge of basic educational principles, including skills for planning, implementing and evaluating teaching and learning. They often have little understanding of the use of teaching aids, or lack a working knowledge of educational methods and their role in the teaching/learning relationship. Not surprisingly, the most common teaching method is the "chalk and talk" approach in a lecture setting.

This book has been prepared for both experienced agricultural teachers and new teachers who are embarking on their careers. It will also be of value to planners, policy-makers and managers of agricultural education. There are many books on teaching and learning that address subjects such as educational psychology and learning theory. Although these subjects are very important to education in general, practical guidelines for teaching agriculture are often missing in standard texts.

In addressing teachers of agriculture who are interested in improving student learning, Abraham Blum combines practical and research experience in teaching and curriculum development with an extensive knowledge of the educational literature and the issues facing education in developing as well as developed countries. Readers will find a broad coverage of problems and proposed solutions related to the teaching and learning process in agriculture as well as references for further study.

By shifting the focus to the learner and the learning process, Dr Blum recognizes the resources which learners bring to the classroom, laboratory and field activities that are part of education in agriculture. Students learn in many different places and ways, and teachers need to take their needs into account when planning courses. Dr Blum's approach to teaching and learning in agriculture aims at bringing out the best in learners by involving them as partners in the education process.

This book which is acknowledged to be one of the projects of the Extension, Education and Communication Service of the Research, Extension and Training Division of the Sustainable Development Department is premised on the belief that sustainable agricultural production and rural development require well-educated young people. There is little doubt that human resource development is an integral part of the national development process. The contents of this book should be useful to both teachers and learners in agriculture who face the challenge of increasing food production and food security in their countries while, at the same time, protecting the natural resource base for future generations.

Stein W. Bie
Director
Research, Extension and Training Division
Sustainable Development Department

Contents

Acknowledgements

The author thanks the Extension, Education and Communication Service for the initiative to write this book and its encouragement throughout the preparation. Special recognition is due to L. Van Crowder, W. Lindley and N. Doron for their significant comments and suggestions.

Permission to use the following figures is gratefully acknowledged.

Figure 3
Source: Foundational Approach to Science Teaching (FAST) Project, Honolulu, Hawaii, 1976.

Figure 4
J. Barker. 1975. Sink or Swim. *London, Science Education Centre, King's College.*

Figure 6
R.W. Bybee. 1991. American Biology Teacher, *53: 146-153.*

Figure 8
K. Frey. 1993. Allgemeine Didaktik, *ETH Zurich.*

All other figures are the author's.

Introduction

ABOUT THIS GUIDE

If you are a practising agricultural teacher or intend to become one, if you work in an educational institution which appreciates the importance of agriculture, or if you just happen to be interested in agricultural education, this guide is intended for you.

Most chapters are relevant for different school levels, agro-ecological environments and socio-economic circumstances but you will find that special emphasis has been put on typical problems and issues faced by agricultural schools and colleges in developing countries. While the focus is on agricultural vocational schools at the intermediate level, much of agricultural pedagogy is also relevant to the teaching of agriculture in "general" secondary and elementary schools.

There are many good books on teaching and learning theories and on their translation into practice. However, these books treat mainly theoretical subjects as they are typically taught in academic streams. As we shall discuss in the next chapter, teaching and learning agriculture encompass much of general didactics and learning skills but they also cover important aspects which are not so relevant or are even absent in the usual classroom situation.

This guide was developed because many agricultural teachers want a book that looks at teaching and learning with special reference to agriculture and that builds on the practical experience of many agricultural teachers and rural educators in a practical form. As most guides, this too can be used mainly in two ways: as a reference for a certain topic on which the more experienced reader hopes to get a useful hint or as a study guide for agricultural education students and the many agricultural graduates who come to teaching without a suitable pedagogical training.

**SPECIAL NEED
FOR EFFECTIVE
TEACHING IN
AGRICULTURE**

Of course, all teaching should be effective. However, we especially emphasize the need for effective teaching in agriculture for several reasons. Agriculture is a complex subject which cuts across many scientific, social and practical disciplines. One of the more difficult tasks of the effective agricultural teacher is to integrate these different aspects in order to give students a holistic view. Especially in developing countries, agricultural school leavers use the knowledge and insights gained not only to improve the farm on which they work but, since they often serve as change agents in advisory services, also to help improve agriculture and the standard of living of farmers in their respective country. To do so, they must not only have much knowledge and many skills in agriculture, they must also learn to understand socio-economic environments and become effective communicators.

Agriculture is the second oldest technology in the world. Only primitive food gathering, hunting and fishing are older. Agriculture was the first human endeavour to interact with nature to *produce*, as well as to collect, food and fibres for the most basic of human needs.

The importance of agriculture has not declined. More than a quarter of the world's population lives in low-income countries in Africa, Asia, Latin America and Oceania. These countries have many things in common; above all, they are poor and have an agrarian economy. Three out of four people work in agriculture but in some countries, mainly in Africa and Central America, they produce less food per caput today than they did ten years ago, in spite of the green revolution. The *First report on the world nutrition situation* (UN [ACC/SCN], 1987) opens with the statement that "hunger and malnutrition cause perhaps the most widespread human suffering in the world today". Children, especially, suffer from undernourishment and unbalanced diets.

In many developing countries, the rate of population growth is higher than the increase in agricultural production. New methods must be found and implemented to raise agricultural production in the future. Droughts make it necessary to keep stores of staple food for times of famine, either in central storage (which is very costly) or on farmers' homesteads. To be able to do this, farmers not only need to know how to grow more food, but also how to prevent pests and diseases from destroying this precious food reserve.

Farmers grow food for the subsistence of their families and for the local market, but also for export, as their countries need foreign currency to pay for imported goods. Agricultural commodities are for many countries, especially in the developing world, the major source of income from exports. However, as the supply of export crops has risen without a parallel increase in demand, the prices they can fetch in food-importing countries have declined. Thus it is necessary to switch over to alternative crops – an endeavour which demands new agricultural knowledge and expertise and their effective spread to the agricultural community.

Agricultural schools are only one component in the agricultural knowledge system, but a very important one. In the past, the oversimplified "transfer of technology" model assumed that practically all agricultural knowledge was generated at research stations and then "extended" to farmers via different communication patterns. Today, we know that the process is not so simple, and a more sophisticated systems approach is needed to optimize agricultural production and use agricultural knowledge as a major production factor.

First of all, an effective feedback from farmers to researchers is needed. By adapting new methods in the field and observing the results, farmers themselves can contribute much to the generation of agricultural expertise. Agricultural knowledge is generated not only by agricultural *experiments*; it is often based on farmers' *experience*, accrued over time. Without this experience, people could not have overcome floods and droughts which have haunted them throughout history. The more effectively experience-based expertise is combined with the ability to adapt research findings to a local situation, the better the chances for improved agricultural performance. It is one of the major tasks of agricultural schools to train students in this integrative skill.

Agricultural knowledge and expertise are transmitted in many forms, through formal courses, informal extension activities, printed and electronic mass media and, above all, through discussions among farmers themselves. However, the agricultural school can play a central role in the agricultural knowledge system. It often trains future extension workers and it should have good contacts with farmers in the area to be able to provide a link between them and research. The agricultural school can serve as a focus for adaptive research and as a demonstration centre. It can be the ideal place to link theory and practice.

TEACHING AND LEARNING AGRICULTURE COMPARED WITH OTHER SUBJECTS	Teaching and learning are universal activities, and therefore much of what we know from teaching and learning theory and from teachers' experience is generally applicable to these activities in agriculture. This is true above all for the basic tasks of all good teaching: the creation of a good learning environment, the fostering of students' motivation and feeling of achievement and training students how to learn and improve their learning skills; and effectively using teaching aids which help to achieve these tasks.

However, students in agricultural schools tend to be

different in some ways from their peers in upper secondary and post-secondary schools. They have chosen agriculture as a career. Not all will have done it out of identification with the rural way of life or because they believe they have a mission to help advance their community. Still, their similar background of a rural environment, together with the ambition to receive at the end of their studies a certificate or diploma in agriculture, can help to create a classroom climate which is conducive to the learning of agriculture as an applied and worthwhile field of study.

The great Greek philosopher Aristotle divided human knowledge into three categories of discipline: those whose task is to know (e.g. mathematics and the basic sciences); those that are "practical" and help to make decisions (among these he included ethics and political sciences!); and, finally, the creative disciplines – among them agriculture and the arts. In modern terms, we can still look at agriculture as a creative activity, but one in which many decisions have to be based on the application of knowledge from both natural sciences, mainly biology and chemistry, and social sciences such as economics. It is this interdisciplinarity of agriculture which makes it so interesting but often also difficult to teach. We shall come back to this point when we discuss the transfer of learning and curriculum development in the next two chapters.

Agriculture is more than a vocational area or even a profession. Unlike many other occupations, it is intrinsically connected with the *land* and working in a *rural* surrounding. Agriculture is also a way of life. It acknowledges a dependency on nature, which it tries to use but which it must also conserve for future generations. More than in many other occupations, farmers share a common fate with their fellow men and women in the community.

Farmers in all parts of the world have created rural cultures. Therefore, agricultural teachers are not just transmitters of specific, examinable knowledge. They should see themselves as educators acting in an environment with which they identify.

BIBLIOGRAPHY

Blum, A. 1988. Science and technology education and agriculture. *In* D. Layton, ed. *Innovations in science and technology education*, p.155-165. Paris, UNESCO.

FAO. 1982. *Training for Agriculture and Rural Development 1981*. FAO Economic and Social Development Series No. 24. Rome.

FAO. 1987. *Report of the Expert Consultation on Progress of Institutional Development in Agricultural Schools and Colleges*. Rome.

Olaitan, S.A. 1984. *Agricultural education in the tropics: methodology for teaching agriculture*. Basingstoke, UK, Macmillan.

Thompson, A.R. 1981. *Education and development in Africa*. London, Macmillan.

UN (ACC/SCN). 1987. *First report on the world nutrition situation*. Rome, FAO.

UN (ACC/SCN). 1989. *Update on the nutrition situation*. Rome, FAO.

UNESCO. 1987. *Higher agricultural education and rural development in developing countries in Asia and in the Pacific*. Bangkok, UNESCO Regional Office.

Wallace, I.R. 1976. Agricultural education and rural development: some approaches and problems in Kenya. *Agric. Prog.*, 51: 131-142.

Teaching
and learning
agriculture

In this chapter we shall consider nine issues which can help to improve the effectiveness of teaching and facilitate students' learning:
- communicating effectively;
- enhancing students' motivation;
- teaching a topic at the right time;
- using advance organizers;
- enhancing the chances for a "transfer of learning";
- teaching principles rather than details;
- learning to apply principles;
- learning to make decisions;
- experiential learning.

COMMUNICATING EFFECTIVELY

Teaching, like other forms of information transmission, is a communication process. Usually the teacher *sends* a verbal *message,* which contains some *information,* to the learners who are expected to *receive* it and integrate it into their existing knowledge.

This process is not so simple. First, teachers have to *encode* their thoughts into words and/or other forms of communication. Then students have to *decode* the message, which means they have to make sense of it. Of course, teachers assume and take steps to assure that what they send is received and decoded by their students in the right way. The clearer the message, the less chance there is of it becoming distorted during the transmission and the easier it is to be decoded by students. To make sure that this actually happens, teachers can strengthen their verbal messages by additional means such as visual teaching aids, thus enabling students to receive the message over two or more parallel communication lines (the ear and the eye). However, the parallel messages must be matched in order to have an amplifying effect. If they are not, they create confusion ("noise", in the language of communication).

Agricultural teachers have an advantage when teaching in the field. Students can observe by themselves and through different channels of perception a situation which the teacher might find difficult to put (encode) into words. On the other hand, being in the field, students are exposed to many more messages (impressions) coming from the environment which can distract them. Therefore, teaching in the field must be as task-oriented as teaching in the classroom (see the sections Teaching on the school farm and Teaching in and with the community, p. 97 and 102).

To make sure that students receive and decode their messages, teachers should look for feedback – a sign that students have understood the message and integrated it into their conceptual framework. This feedback can be in different forms, for example verbal, when students answer questions. Feedback is often encoded in non-verbal signs, for example when students express in body language their reaction to what they have decoded from the message: they nod with their heads, they laugh after a joke, they look bored and so on.

Messages that are received by the students are filtered and stored temporarily in the short-term memory. They are

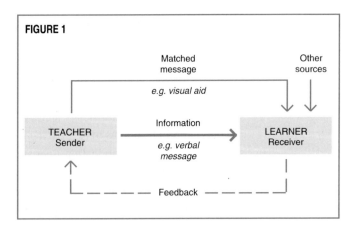

FIGURE 1

forgotten after about 30 seconds if they cannot be kept in mind or transferred to the long-term memory. Thus, we forget casual telephone numbers very quickly unless we make an intellectual effort to remember them. The long-term memory receives new information better when it fits into a framework of concepts which already exists. Incomprehensible and unclear messages are not easily stored in the long-term memory and they are quickly forgotten. Competing verbal and audiovisual messages are difficult to cope with. Showing something to students and talking about something different weakens the transmission of the message. Even a blackboard left over with notes from a different subject can distract students' attention and weaken their reception of the teacher's new message.

ENHANCING STUDENTS' MOTIVATION	• We can make an effort to teach well, but when students are not motivated not much learning will occur. Therefore, successful teachers try to identify the type of motives which activate different groups of students and choose suitable teaching techniques to enhance students' motivation accordingly.

Students' motivation can be *intrinsic* or *extrinsic*. In the first case, students are motivated from within themselves. They *want* to learn the subject or topic taught. They might do so for different reasons. Some students might be driven by intellectual curiosity. The teacher can keep their motivation high by giving them tasks which enable them to discover the answer to open questions and queries by themselves (e.g. from books or by observation). Such students are attracted by challenges such as project work.

Often, the *specific topic* taught arouses the interest and motivation of students. Especially when learning agriculture, students may already have had to tackle a similar problem and therefore see the topic's direct relevance to "real life". These students' motivation is reinforced when they feel that the topic taught is worth learning for its practical value.

Psychology teaches us that the *need of achievement* is a powerful motivating factor in many individuals. Therefore, it is good practice to present such students with tasks at a level at which they can tackle them if they make an effort. Thus, they will feel they have achieved what was expected of them.

Two other types of motivation are closely related to the need of achievement: competence motivation and the novelty of the task. Some students seek satisfaction in achieving competence and mastery in a certain field, especially when this gives them a certain control or enhanced influence over their environment. They are often attracted by the novelty of a task which puts them in a new, unfamiliar or changed situation.

Students with a high need of achievement respond well to a grading system which gives them a high ranking. However, such a system can have very negative effects on pupils with a low need of achievement or with a low self-image.

Learning psychologists have found that, insofar as individuals are permitted to engage in the planning and execution of activities, they become highly motivated. However, this is only possible when students are allowed to say what they think and to criticize suggestions made by the teacher. Such behaviour might be contrary to a school's norm. Where it is acceptable, teachers can use the motive of autonomy to arouse high motivation by arranging situations that involve participation. The motive for autonomy and participation is also powerful for another reason. It is closely related to a variety of other motives, such as a need of recognition of achievement, status seeking, a need of change, curiosity and competence.

However, it is not only the topics of instruction and the tasks put before different students that reinforce (or diminish) their motivation depending on their personality traits. The motive for much human learning resides in the *interpersonal relationship* between teacher and student. This view is based on the finding of many psychologists that acceptance and approval are strong personal needs in most people. Many students tend to respond to the teacher who conveys interest in them as individuals and is ready to listen carefully to them and to respect them as individuals. This is especially important where students' motivation is based more on the *need of affiliation* than on intellectual curiosity and task-orientation.

It is not only the teacher who reinforces or suppresses motivation; interpersonal relationships among students have similar effects. If the classroom is organized to facilitate interaction among friends and work carried out in congenial groups, students tend to be energized and motivated. In a well-functioning group, weaker students learn from their more able peers without hampering the latter. Also, affective goals are often better achieved in a group learning situation where socialization occurs without the repression of personal traits.

At the agricultural school level, the sex drive too can play a motivational role in coeducational classes. The wish to attract favourable attention from members of the opposite sex can reinforce students' motivation to excel. However, the teacher will also be alert to the negative effects which competitive behaviour can also produce.

Praise and reproof as incentives can be cues for both achievement and affiliation motivation. However, praise is the more effective of the two because reproof is felt by some students as rejection.

When intrinsic motivation is missing, students can still do well in their studies if they are motivated *extrinsically*, for example by their will to receive a certificate for its "market value" rather than for the competencies it certifies. Students in this category respond to teachers' comments on their chances of obtaining the desired document. Of course, also negative factors such as the fear of punishment or social disapproval can affect motivation.

It is important for teachers to know their students' expectations. These might be unrealistic and, in such a case, it is better to clarify what students can and cannot expect than to let illusions develop.

Individual students are motivated by different needs and react positively to different motivating techniques. Therefore, it is not easy for teachers to reinforce the motivation of all their students at the optimal level. The most important prerequisite for teachers' success in their important task is to know their students, including their personality traits and expectations. Therefore, it is in many cases better for teachers to teach several courses to a given class, and thus become better acquainted with their students, than to concentrate on a specialized subject. This is often better than meeting a class for only a few hours per week and then rushing to another class.

Among the different teaching approaches, guided discovery or inquiry teaching has an especially high motivational quality. On the other hand, it is a more time-consuming technique (it is discussed in more detail in the section Teaching styles, p. 74).

TEACHING A TOPIC AT THE RIGHT TIME

Bruner (1960) is said to have regretted his famous, provocative hypothesis that "any subject can be taught effectively in some intellectually honest form to any child at any stage of development". He did so because this sweeping statement cannot be used by teachers as it stands. However, it reminds us that teachers have to take into account the stage of development and previous learning experiences of their students when deciding what, when and how to teach a certain topic.

Piaget and Inhelder (1969) and many researchers after them have studied the intellectual development of children as they progress from one developmental stage to another. At the critical age of adolescence, students move from the *concrete-operational* to the *formal-operational* stage. At the *concrete-operational* stage, students develop an internalized, conceptual structure for the things they encounter, but they are not yet able to deal with possibilities not directly before them or not already experienced. They cannot go systematically beyond the information given to them. When students pass into the *formal-operational* stage, they are able to operate on hypothetical propositions. They are no longer constrained to what they have experienced or what is before them. They can now think of possible variables and even potential relationships which can later be verified by experiment or observation. At this stage, students can express their thoughts in abstract terms without needing to refer to concrete events.

Several factors influence the intellectual development of individuals. Genetic as well as environmental factors influence the transition from the concrete to the formal stage. Teachers can help students to pass progressively from concrete thinking to the utilization of more conceptually adequate modes of thought. The problem for practising teachers lies in the difficulty of knowing where each of their students is in this developmental process, as there are no ready tests available for this purpose. Based on their age, students in agricultural schools should be at the formal-operational stage, but the teacher cannot assume this to be true for all students. Actually, many adults never reach the full level of formal thinking. However, the teacher can use several techniques to be effective also with students at a lower level of concrete thinking.

One of the advantages of teaching agriculture is that it

deals mainly with concrete, real life situations. When we want to come to generalizations (the value of which is discussed later), we must work at a higher level of formal thinking. Yet we can bring students who are only at the threshold of formal thinking to grasp generalizations at an "instrumental" level, as illustrated by the following example.

A group of students carried out a controlled experiment to find out how plastic tunnels affect the rate of growth of vegetables. When asked at the end of the experiment why they had used a control plot, students at the formal stage of their intellectual development answered with a more or less correct definition of a control as it would appear in a dictionary, i.e. "the control is the part of an experiment in which the procedure or agent under investigation is omitted, and which is used as a standard of comparison for judging the experimental effect". What a lot of abstract terms and concepts there are in this sentence! Students who were still at a rather concrete level of operation avoided the formal definition (which they might have been asked to learn by heart, but obviously did not digest). Referring to the concrete experiment they had conducted, they said that they "had left one plot of vegetables without plastic tunnels on purpose so that they could compare these vegetables with those growing under plastic", but added that they "made sure that all the other things they did in the two plots were exactly the same, otherwise they would not know what had caused the difference". The test for this *working understanding* came when these students were asked to plan an experiment to find out how a change of day length would affect the time of flowering of chrysanthemums. They were perfectly able to propose the right control.

This example also shows the use of the *black box* approach, which is useful when we want to teach how to use an agricultural technology before students are able to understand the scientific basis on which the technology is based. Actually, we use "black boxes" all the time. The television set is one. Who knows what is really going on inside this box? Yet we know which buttons to push. In the first experiment described above, the teacher did not discuss the different physical factors which might have caused the quicker development of vegetables under plastic (a higher temperature and CO_2 concentration in the early morning, less wind and possibly additional factors). In the next experiment, the teacher could not explain the effect of infrared light and the role of phytochrome in the flowering induction process. Had the students had a better background in biochemistry and physiology, the teacher could have given a more complete explanation of what happens in the plant. But, for practical purposes, the level at which the lesson was taught was sufficient.

These cases show that we can teach the same topic at different levels in "an intellectually honest form", as Bruner (1960) argued. However, we have to be careful not to create misconceptions which are difficult to correct later. Many students have difficulties in understanding the law of energy conservation after they have been taught erroneously about chemical processes in which "energy is lost".

The optimal solution is that of a *spiral curriculum* in which major topics are treated at different stages in the curriculum, but each time additional information is supplied according to the prerequisite scientific knowledge which students have acquired in the meantime and which is needed to understand the new and more abstract information. Thus, the older information is reinforced and serves as an "advance organizer" (discussed in the next section).

Part of the problem of "when to teach what" is the wrong interpretation of Auguste Comte's *hierarchy of disciplines*. The French natural philosopher ordered the disciplines of the sciences according to a hierarchy in which the findings of each discipline can be described in terms of a discipline at a more basic level. At the base is mathematics, as a kind of natural logic, in terms of which the findings of physics can be described and put into mathematical form.

Findings in chemistry can then be reduced to physical principles; the characteristics of biological organisms can be seen as complex physico-chemical systems; psychological characteristics can be expressed in biological terms; and even sociological phenomena can be conceptualized as aggregates of psychological systems.

One could also add agriculture at the top of the pyramid (see Fig. 2). Universities tend to base their curricula for agricultural studies on this hierarchy, demanding that students first master mathematics to be able to understand the physics and chemistry courses which, in turn, are considered a preparation for the understanding of physiology. The subject students have chosen to learn (agriculture) is postponed for a long time, thus students' intrinsic motivation is lost.

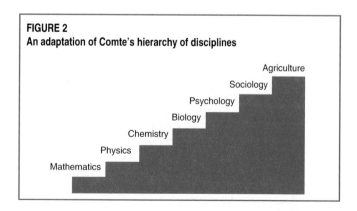

FIGURE 2
An adaptation of Comte's hierarchy of disciplines

Agriculture
Sociology
Psychology
Biology
Chemistry
Physics
Mathematics

In an agricultural school, the teaching of agriculture cannot be postponed until students have mastered biology and, before that, chemistry, physics and mathematics. Therefore, agricultural teachers must do two things: i) As already mentioned, when teaching a topic, they are often obliged to leave some of the more basic explanations to a later stage in a spiral curriculum because they cannot wait to postpone all topics until students have optimally mastered all the prerequisites. In this case, they will treat some items as "black boxes". ii) When teaching applied agricultural topics, teachers will try to integrate scientific explanations at a level at which students can understand them.

USING ADVANCE ORGANIZERS	"Advance organizer" is a term coined by Ausubel (1963) to describe an important psychological principle in learning and teaching. Single facts are difficult to learn out of context and are quickly forgotten. When we can attach new pieces of knowledge to already existing concepts or even whole conceptual structures, they are learned and retained better. This can be compared with a file ordering system in which the drawers and files are already well marked; a new document can easily be deposited in the right file and drawer and can also be readily retrieved.

Advance organizers can be used by teachers in different ways. The more clearly a teaching episode is constructed, the easier it is for teachers to present students with what they are about to teach and to explain how it relates to previous learning. When students are asked to learn from written sources such as professional publications, they should be instructed to pay special attention to the introduction and summary in papers and to headings used in chapters or an essay, instead of going directly to the material itself. Experienced readers and researchers do the same. Abstracts at the beginning of a chapter or paper can serve as good advance organizers and facilitate an understanding of the main part of a paper.

ENHANCING THE TRANSFER OF LEARNING

How much does learning in one area serve to improve the learner's performance in other areas? Or, put in a different form: how adequately does training in one situation generalize to other situations? These questions are especially important for agricultural teachers who, in an applied course, often rely on understandings and insights gained by students in an earlier basic course. Thus we can ask, for instance, how far will instruction in soil chemistry and plant physiology influence a learner's ability to plan crop fertilization?

Educational research has yielded some generalizations which can help the teacher to enhance the chances for a transfer of learning to occur. However, teachers should not automatically rely on its occurrence in a specific case. Many factors influence the degree to which learning is transferred. The most important are described in the following paragraphs. The reader will observe that some psychological and instructional principles which have been found to be useful in earlier sections will reappear in the context of the transfer of learning.

Not all students have the same capability to transfer learning from one area to another. Age, mental ability, attitude towards learning and acceptance of the method of instruction have been found to influence the transfer of learning. Older and intellectually brighter students transfer their learning more readily than their younger or less intelligent peers. Especially important for teaching agriculture is the fact that when students regard what they have learned as being useful beyond the classroom, the transfer of learning is enhanced. This is another reason to integrate practical examples in the teaching of agricultural principles.

The transfer of learning is also improved when students *learn broad concepts and principles* which are applicable in different situations, and not just facts. Good teachers give their students examples from different possible areas of *application* when they explain important generalizations. This is especially useful when basic principles of biology are discussed. Thus, the curve of diminishing growth (which is also the curve of diminishing economic returns) can be taught with a number of useful applications, for example the optimal amounts of chemical fertilizer to be spread, of irrigation water to be supplied or of concentrates to be fed. The greater the number of similarities the student perceives between the original learning and other situations in life, the more the original learning can improve performance in other situations.

A transfer of learning is more likely to occur when the two *situations are similar* and when the new situation occurs *shortly after* the knowledge to be transferred is learned. Also, directions given by the teacher can enhance students' chances of being able to transfer learning effectively. Teachers can provide students with a diversity of problems in which they practise the application of newly learned skills and principles to varied life situations, for example situations which typically occur on farms. These problems can be at different levels of similarity:

i) similarities which are quite apparent;
ii) similarities which exist but which are not readily identified by students; and
iii) similarities which exist but which are not really useful.

Experience has shown that the intermediate level is the

best to use because it demands an intellectual effort from students while also giving them the satisfaction of discovery.

Concepts and generalizations which the learner derives from *investing personal efforts* (gathering data from different sources, drawing conclusions) transfer better than those which the student was taught in the form of verbal definitions. When students have to find a solution to a problem by themselves or with only partial guidance, the transfer of learning can be expected to be better than when students learn passively by listening to a lecture or even observing a demonstration by the teacher.

The transfer of *technical skills* seems to be more restricted. In most cases it was found that, with practice, the speed and quality of a given technical task can be improved but that this does not help to improve other practices. However, the transfer of practical training can be enhanced to some extent when students understand the principles which underlie the practices. In agriculture, this means that we can enhance the teaching of practices when we make sure that students understand *why* they should do things the way they are taught. For instance, a student who has understood that in cleft grafting the most important thing is to bring the cambium of scion and stock into close contact, and who has got used to doing this, will adapt to the parallel procedure in whip grafting more quickly than a student who has not received training in a previous type of grafting and who does not understand *why* the two tissues should be joined closely together.

TEACHING PRINCIPLES RATHER THAN DETAILS	In the last sections, we have come across the advantage of teaching principles in several contexts, for example by using them as advance organizers or to facilitate a transfer of learning by building on the application of principles in

new situations. Emphasizing principles in teaching has an additional advantage. The amount of knowledge in all areas, but especially in the sciences and technologies (to which agriculture belongs), is growing at an exponential rate. Based on the number of scientific papers published, the amount of knowledge is doubled about every ten years. Thus, even if students (or teachers) acquired all the knowledge relevant to a certain field of study and did not forget anything (two impossible suppositions), after 20 years, when they should be at the peak of experience, they would only have a quarter of the knowledge which would have accrued in the meantime – and a large part of their former knowledge would certainly no longer be up to date or correct.

What changes quickly are detailed pieces of knowledge. Basic rules, principles and generalizations in all the sciences change much less. In the late 1950s, a group of biologists listed the most important principles of their discipline, and this list has not changed since.

Furthermore, in a study done on students' retention of learning (Tyler, 1933) it was shown that, within a year, students had forgotten 77 percent of the specific facts they had learned (the names of animal structures in a diagram). At the same time, the students' ability to apply a principle to a new situation was unchanged and the skill to interpret new experiments even improved by 25 percent, probably owing to the additional experience gained in using this intellectual skill.

How, then, can we cope with the quick loss of specific facts from our memory (when they are not used after learning)? Probably the best answer is to concentrate on the teaching of principles and to use specific facts mainly to demonstrate how the principles work. Of course, specific facts can be most important – once we need them.

However, instead of letting students learn these specifics by rote (which is not conducive to long-term retention), it would be better to teach students how to find the details, if and when they need them. Thus, practice in the use of dictionaries, technical handbooks, agricultural compendiums and extension publications becomes an important educational goal.

Remedying the loss of details in our memory has something in common with solving the problem of "knowledge inflation". In both cases, it is important to know where to find the most updated information in a suitable form.

Furthermore, it is imperative that students *learn how to learn* by themselves – because they will have to do this for the rest of their working life if they do not want to fall behind younger and more up-to-date colleagues. Life-long learning is not only a skill. It is also an attitude and habit which is acquired over time, mainly by exercise, and which should begin in school.

LEARNING TO APPLY PRINCIPLES

In Bloom (1956), the term "application" as a mental skill stands above "knowledge" and "comprehension" because only a piece of knowledge (e.g. a principle) which has been comprehended by students (and not only learned by rote) can be applied to a new situation. The fact that most of what we learn, especially in agriculture, is intended for application to problem situations in real life is indicative for the importance of application objectives in the curriculum and of training students in applying principles. Much of what was discussed under the transfer of learning has to do with the application of principles. Research studies have shown that comprehending a generalization does not ensure that an individual will be able to apply it correctly in a new situation. Training is needed to develop the skill and ability to apply generalizations in problem-solving situations.

Real life problems can be quite complex, because many factors have to be taken into account and the problem might be very different from what the problem solver has experienced before that. Research has shown that two aspects define the difficulty of applying principles, rules or generalizations to the solution of a problem:

  i) the number of principles one has to think of when trying to solve the problem; and

  ii) the remoteness of the new problem situation from the situation in which the principles were studied.

The complexity of the problem-solving situation depends on the number of principles to be considered. This explains why real life agricultural problems, with their many natural, economic and social factors, are often so complex and why training agricultural students in problem solving is so important in the teaching of agriculture.

When devising application exercises, the teacher might start with a situation which is close to the one the student has just learned. In a soil science course, the problem situation would also be in soil science. For instance, after students have learned about the affinity of cations to negatively charged clay particles, they can be asked how this principle will affect various ionized nutrients in the soil. The next step could be to pose the question of how different chemical fertilizers will behave in a clay soil. In this case, students have to consider not only the chemical composition of the fertilizers but also their degree of ionization (which they might have to find in a source book or separate list provided by the teacher). The harder task will come later, when students might be requested to develop an annual fertilization plan, where the issue of fixation is just one of many principles to be considered.

Graduates of agricultural schools will often work in different parts of their country. After they have studied a farm problem on the school site and have come to a conclusion, they can then be challenged to propose a solution to the same problem, but assuming that it arises under different agro-ecological conditions (which would either be supplied by the teacher or found by the students in relevant written sources).

In a test on students' ability to apply the rules of agricultural experimentation (Blum, 1989), students were asked three sets of questions. In the first set of questions, they had to draw a conclusion from the results of an experiment. To answer correctly, only *one* rule had to be applied, namely that results of an experiment are valid only for the population represented by the sample. In the second set of questions, students had to validate the correctness of an experimental design. To do this correctly, they had to apply *three* experimental principles that they had learned: controls, repetitions and randomized samples. In the last set of questions, students had to solve an agricultural problem by applying the correct biological principle chosen from an *unlimited* number of principles, of which they had learned quite a number.

The results showed clearly that these three sets of questions were at different levels of difficulty owing to the growing number of principles involved and their greater complexity. Especially in the last set, in which biological principles had to be applied to the solution of an agricultural problem, there was a clear difference in students' ability to solve questions relating to a familiar situation and questions describing a different, new situation.

LEARNING TO MAKE DECISIONS

There is a big difference between problem solving in a school situation and decision making in real life situations. When problems are posed in a typical learning situation – even when problem solving is taught with the help of application exercises – usually only *one* solution is expected, i.e. "the right one". When we have to make decisions, we usually have to choose among several alternatives. For each alternative, we have different amounts of information available. We are influenced by the sources of information and how they are qualified. We have to consider costs and possible positive and negative outcomes. Above all, we are influenced by our own value system and preferences.

Consequently, it is difficult to approach decision making in a systematic way. Yet, the educational literature is full of statements on the paramount importance of teaching in school, and especially at a vocational level, how to improve decision making. When a group of directors of applied science curriculum projects were asked what they thought about the need to introduce decision making into school curricula, nearly all gave a positive answer. However, when further questioned as to whether they had developed an appropriate decision-making exercise or planned to do so, only a few could answer in the affirmative.

Here, agricultural schools and colleges with a training farm attached have a clear advantage. For learning how to make decisions realistically certain conditions have to be met:

- A real issue should be at stake, about which a certain degree of disagreement exists among students.
- Students must be provided with the body of information necessary for making the decision, or should know how to acquire this information.
- Students must feel that they are able, entitled and expected to express opinions and reach solutions.

• Students should have the possibility of carrying out their decisions so that they can evaluate them and draw conclusions from this evaluation.

Techniques

Part of the problem is that decision making as a technique has been developed mainly by economists, using complicated mathematical models which are not suitable for "home use". On the other hand, many believe that decision making is more of an art than a science. However, "artistic behaviour" can also be taught, and different techniques have been developed to train students in decision making. Following are some different approaches:

Decision trees lead the learners systematically through a series of decision points. At each point, they have to consider the costs of two alternatives and weigh them against the expected gain. Figure 3 gives an example in which students had to weigh the advantages and disadvantages of intensive versus biological farming. Mainly quantifiable factors are taken in account in decision trees.

Cost-benefit analysis. Real decision making also involves monetary factors. What will happen if a certain action is *not* undertaken? What will the social cost of a certain decision be? What qualitative gain (e.g. satisfaction) could be the benefit of an alternative decision? In this type of cost-benefit analysis, students receive a brief description of an agricultural situation in which a decision has to be taken. For each alternative they are asked to list the positive and negative outcomes and to justify why they would choose one of the alternatives or suggest a compromise between two options.

Flow chart improvements. Students are presented with a flow chart in which a valuable resource (labour, water, foreign exchange, etc.) should be economized on. Our example (Fig. 4, p. 32) shows the flow of water in a potato chip factory, set up to bring more income to a potato farming community which also had to economize on water. Students are challenged to find ways of saving water in the production process.

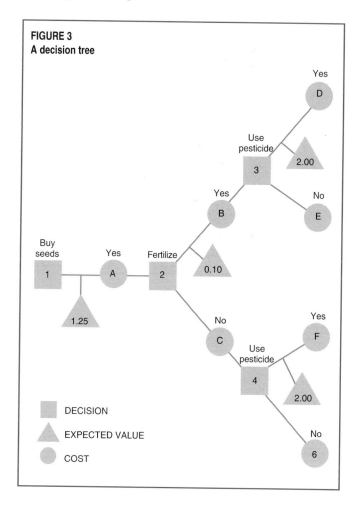

FIGURE 3
A decision tree

Case-studies with implications. These are usually accounts about controversial issues of a more general character, for instance cases in which developmental advantages have to be weighed against environmental hazards. Examples could be a project in which it is planned to clear bush and convert it into arable land or one in which an irrigation dam is planned, but where the cost could be possible ecological damage or the displacement of local farmers. Again, students are asked to analyse the situation and to take sides or come up with a balanced solution based on the facts at hand and their own value judgements.

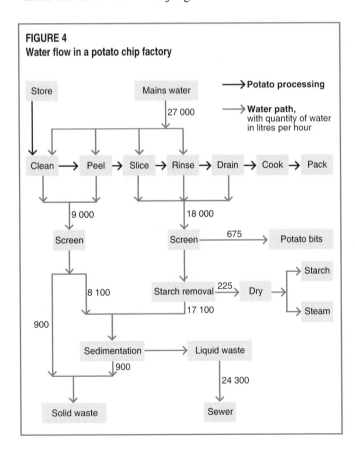

FIGURE 4
Water flow in a potato chip factory

Role play and simulation games can be used to structure the handling of the case-studies and to make the tasks more lively. These two learning/teaching methods are discussed in more detail in the section Exercise games, role play and simulation games, p.131.

EXPERIENTIAL LEARNING

Recently, the "experiential learning" approach has been developed for teaching agriculture. This approach views learning and the farming environment as "soft systems", meaning systems that are not fixed. When one tries to separate the elements of a system in order to study each element in isolation, something important is lost, even if an attempt is made later to reintegrate the parts. Therefore, the "wholeness" of a system (e.g. the agricultural environment as a whole) should be studied. Without looking at the whole, the parts cannot be fully understood.

Experiential learning is not only based on school knowledge and acquired skills, but also on *experience*; hence its name. Experiential learning is practised by many people who have never gone to school but who grasp what is going on around them and creatively take action to adapt to constantly changing situations. It is a combination of "finding out" and "taking action". The process involves feelings, attitudes and values which markedly affect the disposition of the learner. These factors are to be found in any decision-making situation, although their importance is not always acknowledged. The idea of experiential learning is close to Freire's concept of "empowerment of the learner in action" (Freire, 1972).

Experiential learning is geared towards problem solving. It is not easy because, in the natural sciences, we are used to the reductionist model of "classical" research whereby factors are separated for better study. Experiential

Experiential learners should develop four basic steps:

- immerse themselves as fully as they can in all aspects of the problematic situation they are encountering and observe what might be happening from as many perspectives as possible;
- "shape" these experiences into meaningful interpretations so that some understanding can be revealed;
- use this understanding to design plans of action to improve the original situation;
- execute the plans in a way that will lead to desirable and feasible changes.

learning in agriculture was developed mainly at Hawkesbury College in Australia. An evaluation of the college's experience showed that graduates were more employable and that their employers believed that their approach to problems was more open-minded and comprehensive, thus enabling them to be good problem solvers and communicators, to be more creative and to be experienced in team work.

BIBLIOGRAPHY   **Ausubel, D.P.** 1963. *The psychology of meaningful verbal learning. An introduction to school learning.* New York, Grune & Stratton.

Bawden, R.J. 1992. Systems approaches to agricultural development: the Hawkesbury experience. *Agric. Syst.,* 40: 153-176.

Bligh, D. 1986. *Teach thinking by discussion.* Guildford, UK, Society for Research in Higher Education.

Bloom, B.S., ed. 1956. *Taxonomy of educational objectives. Handbook I: cognitive domain.* New York, McKay.

Blum, A. 1989. Facet theory used to establish taxonomic subclasses for "application", with an example from science teaching. *Sci. Educ.*, 73: 1-12.

Bruner, J.S. 1960. *The process of education.* New York, Vintage Books.

Child, D. 1981. *Psychology and the teacher*, 3rd ed. London, Holt, Rinehart & Winston.

Davis, M.D. 1985. *The art of decision making.* New York, Springer.

Dreyfus, A. 1985. Education for decision making in agriculture: scientific concepts in authentic situations. *In* G.B. Harrison, ed. *World trends in science and technology education*, p. 172-174. Nottingham, UK, Trent Polytechnic.

Fox, M. 1993. *Psychological perspectives in education.* London, Cassell.

Freire, P. 1972. *Pedagogy of the oppressed.* Harmondsworth, UK, Penguin.

Gagne, R.M. 1974. *Essentials of learning for instruction.* Hinsdale, Ill., USA, Dryden.

Good, T.L. & Brophy, J.E. 1990. *Educational psychology – a realistic approach.* London, Longman.

Law, B. 1977. *Decide for yourself.* Cambridge, UK, Hobsons.

Lesgold, A. & Glaser, R., eds. 1989. *Foundations for a psychology of education.* Hillsdale, N.J., USA, Lawrence Erlbaum Associates.

MacMillan, P. & Powell, L. 1981. *Perception and learning. An induction course for teaching in further education and industry.* Huntingdon, UK, Transart.

Novak, J.D. & Gowin, B.D. 1984. *Learning how to learn.* Cambridge, UK, Cambridge University Press.

Piaget, J. & Inhelder, B. 1969. *The psychology of the child.* London, Routledge & Kegan.

Pye, J. 1988. *Invisible children: who are the real losers at school?* Oxford, UK, Oxford University Press.

Sutherland, P. 1992. *Cognitive development today: Piaget and his critics*. London, Chapman.

Teacher Training Section, Dronten College of Agriculture & Pedagogical Advisory Center Amsterdam. 1982. *Practical guidelines for effective teaching and learning in agricultural education and training*. The Hague, Ministry of Agriculture and Fisheries.

Tyler, R. 1933. Permanence of learning. *J. Higher Educ.*, 4: 203-204.

Curriculum
development

A differentiation is sometimes made between the *syllabus,* which contains an outline of the educational objectives, the contents and methods of a subject or course, and the (full) *curriculum,* which includes learning and teaching materials as well as the syllabus. However, more often, only the term curriculum is used, and we shall follow this usage.

Practically all countries have curriculum guides which are developed at a national or local level. Some are more centralistic, especially when they are geared towards national examinations. Educational authorities in most countries realize that central curricula can only be a guide, and that a realistic school curriculum must take into account local conditions, the composition of the student population and the facilities available, but also teachers' abilities, training and inclinations. After all, it is the teacher and not the curriculum guide who interacts with students. This does not negate the importance of systemwide curriculum guides. However, here we shall look at the curriculum from the point of view of a school. The basic types of questions asked are similar for both national and school curricula. In the first section, some basic principles of curriculum development are discussed. The following two sections, Agriculture in the school curriculum and Coordinating and integrating subjects, deal with issues which are typical of or even specific to agricultural curricula, while the last section treats the final step in curriculum development: lesson planning by the teacher.

BASIC PRINCIPLES OF CURRICULUM DEVELOPMENT

The most influential model for curriculum development is that of Tyler (1949) (see Fig. 5, p. 41). Although published more than 40 years ago, it has not lost anything of its common sense approach, and it lends itself well to agricultural schools. The model is based on three interconnected focal points: objectives; learning experiences and their organization; and evaluation. Usually, one begins with objectives and, on the basis of these, develops learning activities. An evaluation will then show if the objectives have been achieved. The results of an evaluation may lead to a modification of the learning experiences or of the objectives (if these are shown to be unrealistic). However, sometimes a new curriculum starts with the evaluation of its forerunner, or even with a set of successful learning activities.

The term "objective" can be distinguished from the more general terms "goal" and "aim". The latter two express the good intentions of the educational system more than concrete definitions of what is expected to be the result of teaching certain instructional units. Goals such as "educating students to become good citizens" or "training students to become successful farmers" express the philosophy of an educational system, but they cannot be easily turned into learning activities and the teacher has no possibility of assessing whether they will be achieved once the student has left school and has become a farmer and/or a full-fledged citizen. More precisely formulated objectives such as "students will be able to draft a graph from data supplied in the form of a table" or "students will be able to keep records on a crop demonstration plot" can be evaluated, and students' progress can be monitored so that the teacher can decide if more exercises or remedial activities are needed to achieve the objective.

The first task in curriculum development, also at the school level, is to define objectives while *considering the needs and constraints of the learners, society and the subject matter*. Often, nationally prescribed or suggested objectives exist already. However, these too must be adapted to the local situation. The expectations of students and of society (in the form of government ordinances, parents' attitudes, generally accepted values or cultural patterns) and of the subject matter (often represented by subject matter specialists and examination boards) are important factors to be considered when objectives are adapted to the special situation of an agricultural school. It goes without saying that the idiosyncratic ideas of teachers and much of their experience also end up in the curriculum they teach (see Fig. 5).

Perhaps the most serious societal issue that affects curriculum development in agricultural schools is the image of agriculture itself. In industrialized countries, agricultural schools are clearly for those who have decided to make a career in agriculture. These students are either the children of farmers who intend to continue farming or who wish to attend the agricultural school as a step towards earning higher qualifications in agriculture, for instance as college-trained agronomists or leaders in a professional agricultural organization. In these countries, agricultural school curricula undergo changes to enable their graduates

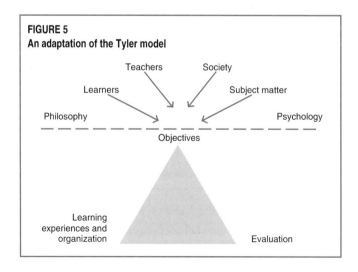

FIGURE 5
An adaptation of the Tyler model

to continue and upgrade their formal agricultural education. They do this mainly by making the curriculum more academic and setting high examination standards. The general atmosphere in these schools is one of rural pride and identification with the farming community.

In many developing countries the situation is different. Because of the considerable socio-economic disparity between farmers and white-collar workers in cities, education is often seen as a way to escape agriculture and rural drudgery. Parents hope that their children will benefit enough from education to achieve the miracle of being able to leave farming. Therefore, part of the curriculum deliberation, especially at the stage of formulating the objectives, should be devoted to the problem of how to use the teaching of agriculture to improve the image of agriculture in the eyes of students. Objectives in the *affective domain* are no less important than the more obvious and common objectives in the *cognitive* and *practical skills domains.* Students' interest, motivation and social skills should be developed parallel to cognitive teaching.

According to the Tyler model, the objectives should be screened to ascertain that they are consistent with both the *philosophy of the school* and what we know of the *psychology of learning* (see the chapter, Teaching and learning agriculture).

Formulating objectives

Objectives should be stated in a way that makes them useful in selecting and creating appropriate learning experiences. It is a good rule of thumb to begin the description of an objective with the words "the student will ..." and then go on to what mental and technical skills the student is expected to possess at the end of the learning period. The most important thing is what the *student learns* and not what the teacher teaches. This is also the reason for speaking of learning *experiences* and not learning activities; when activities planned by the teacher do not induce a positive experience in students, their impact will be short-lived.

Educational objectives can be formulated at different levels of specification. The basic level describes the *performance* expected from students after they have learned a task; for example, *"students will be able to identify the most common weeds"*, or *"students will be able to conduct an extension meeting with a group of subsistence farmers"*.

Sometimes, *conditions* under which students are expected to be able to perform the task are added. Thus, to the two objectives one could add the following conditions: "students will be able to identify the most common weeds, *given a tray of grasses* (or, *while in a cornfield*)"; and "students will be able to conduct an extension meeting with a group of subsistence farmers, *with whom they are familiar*".

At a third level, *standards of performance* can be set which quantify the expected achievement. Where quantification is relatively easy, these standards can be used mainly in examination situations as an objective measure. Thus, one might demand that students be able to identify *20* common weeds, given a tray of grasses (or in a cornfield), and even specify that they *identify* [*them*] *within 15 minutes, not making more than two mistakes*. This degree of detail is less important for teachers but would make it clear to students how they are expected to perform in an examination. Setting standards of performance is more problematic for qualitative objectives such as that of conducting an extension meeting with subsistence farmers. A standard could be added, for example *"to the satisfaction of the audience"*, but how can this be tested in a realistic situation?

Learning experiences

Learning experiences are built into clearly defined learning *episodes*. These can be of varying duration: from a few minutes to the major part of a lesson. Each episode concentrates on the learning of a specific objective by a suitable method. Of course, it cannot be expected that an objective will be reached in just one learning episode. Many more will be needed to achieve an objective, but devising different learning episodes based on the use of different teaching methods and techniques enhances the chances of students mastering an objective.

By using different teaching methods and techniques in a series of learning episodes, the teacher can reach more students because they differ in the way they react to various teaching methods. Also, preparing short learning episodes and using different methods helps to keep the attention of students, especially in the case of intellectually weaker ones.

Learning episodes should be organized in a sequence where each builds on the previous one or on other previous learning experiences. Thus, one episode in a series reinforces existing knowledge and skills and it, in turn, is used as an advance organizer for the next episode.

However, a series of episodes planned to serve a given objective does not always occur as a chain, with each episode being taught one after the other. In a *spiral curriculum* the teacher might come back to treat a topic or objective again at a later stage, after students have learned certain prerequisites – perhaps from another subject – which are needed to treat the topic at a higher level.

Curriculum planning is not complete without *evaluation*. This serves not only to assess students' performance in respect to an objective or standard (as detailed in the chapter, Monitoring and evaluating students' learning), but also to improve, update and revise the curriculum on a regular basis.

Curriculum development is a demanding job. Many aspects have to be taken into account besides the subject matter. Therefore, the best curriculum development jobs are achieved by *teams of teachers.*

Agriculture is closely connected with other subjects in both the natural and the social sciences (e.g. economics) and serves also as a vehicle for value education. Therefore, teachers from these subject matter areas should be called in when issues arise that overlap with other areas.

AGRICULTURE IN THE SCHOOL CURRICULUM	**The agricultural-vocational curriculum**

The agricultural-vocational curriculum

Agricultural schools at the intermediate level differ in the amount of time they devote to agriculture but vocational agriculture plays a central role in the school curriculum. It should be planned in accordance with the objectives of the technical-vocational education strategy of the country and the projected needs for human resources in agriculture.

This needs assessment is not easy. In many developing countries, most of the school's graduates will work in a state extension service or in another governmental position, for which job profiles are usually available. However, during the last decade the employment market has changed in many countries.

More and more graduates of agricultural schools and colleges are finding employment in the private sector, not only in production agriculture but also in the agricultural service sectors (e.g. supplying production means, marketing, food handling). Where suitable conditions exist, for instance when irrigation projects enable new farming units to be set up, the number of graduates who will go on to be self-employed will rise. In industrialized countries, many agricultural students at the intermediate level come from farms and go back to them, assuming they will one day take over the family farm. A similar development can also be expected in developing countries, once the level of sophistication has risen, especially in agricultural commodities sectors which have to compete on the world market. Therefore, a balance has to be found between the currently envisaged occupations students will take up shortly after leaving the school, and those projected for 10 to 15 years later when the former students will be at the height of their professional development. While immediate training needs and the required qualifications must be analysed, the curriculum base should be wide enough to stress basic needs that will outlive structural changes occurring over the next two decades.

Special emphasis should be put on a thorough knowledge of the local situation because this is the environment in which students will learn. This knowledge should be based on an analysis of both agricultural aspects (e.g. farming systems under rain-fed and irrigated conditions) as well as socio-economic and socio-cultural aspects (e.g. categories of farmers, their social and cultural affiliations). However, the curriculum should also take into account the fact that many students will work in other parts of the country. Therefore, they must be able not only to understand their present environment but also to analyse by themselves the situation in another region (to which they might be posted).

The assumption that most graduates will become change agents in one form or another makes it necessary to train students also to be adult educators and successful communicators.

Additional roles of agricultural education

Besides agricultural-vocational education, which is the main focus of this guide, agriculture can play additional educational roles of importance in schools, either as a separate entity or in relation to other subjects.

Historically, agricultural education in schools began with gardens in which students learned to work and appreciate the product of their efforts, and where they could study living plants. These goals are still felt to be important in many places, but they have also changed over time. The following paragraphs describe the major trends in non-vocational agriculture.

Pre-vocational agriculture (as career education)

In both industrialized and agrarian countries, agriculture is offered in elementary and junior high (or lower secondary) schools with the aim of familiarizing students with agriculture and giving them an opportunity to appreciate it. The first aim dominates in industrialized countries where many students no longer know how their food is produced or what farmers actually do. By growing plants and caring for animals, students can decide better if working in the fresh air with living organisms appeals to them enough to choose agriculture as a profession. In some countries, students in intermediate school are also exposed to other vocations to help them decide what direction to follow in their further studies or in their career choice. In this context, agriculture is mainly offered in urban surroundings where students have no first-hand experience with it.

The main problem with school gardens for pre-vocational education is their inability to give a true picture of modern, progressive agriculture. Even agricultural schools find it difficult to be up to date with the quick modernization and sophistication of commercial farms, much more so school gardens. Therefore, pre-vocational agricultural education should include visits to successful farms (not necessarily large ones) and talks with their managers.

Agriculture as a hobby

School gardens and school farms have always produced some students who enjoy working with plants and animals but who would not choose agriculture as a career. However, they may show enough interest in it to develop it into a lifelong hobby. For this purpose, agricultural teachers have offered classes or founded clubs for youngsters (and in some cases also for adults) who want to obtain the skills required for farming small plots near their home or in community gardens, or for growing special home plants (e.g. cacti or orchids). In some countries this is done under the name of rural studies, with a strong emphasis on giving people in an industrialized society an opportunity to create an emotional balance between the hectic city life and the enjoyment of having contact with plants and animals.

Agriculture as compensatory education

Agricultural programmes are sometimes chosen for their compensatory value for mentally and emotionally disadvantaged children. While these programmes have benefited many people, they have also contributed to the dubious notion that agricultural education should be offered mainly to underachievers and the non-motivated. This is mainly the case when mentally retarded adults are allowed to work on school farms out of humanitarian considerations, and students draw the wrong conclusions.

Appreciation of agriculture (as part of development education)

In some agrarian countries agriculture has been introduced into the elementary school curriculum because the educational authorities are concerned about the negative image of agriculture and the migration of rural people to urban centres where they have difficulty in finding work.

Thus, agricultural education is seen as part of development education to demonstrate the importance of agriculture for feeding the population and for export. In most countries this has not worked well because the curriculum has been too technical and has not focused on the main purpose, or because teachers have not identified fully or in a persuasive fashion with agriculture.

Some industrialized countries have also included elements of agriculture in their school curricula. This has been done partly for similar reasons – to give a more positive image of agriculture, with which students are not familiar. In these cases, the curriculum has emphasized the scientific character of modern agriculture and (where appropriate) its contribution to exports.

In other industrialized countries, curricula (often in social sciences) serve to make students aware of problems in agrarian developing countries. These curricula treat problems such as inequality between urban and rural areas, population problems, food calamities and food security.

Farms to teach biological phenomena

School gardens and larger school farms have always been used to give students an opportunity to study live plants and animals in various seasons and stages of development. The best way to do this is by letting these plants and animals grow naturally, or at least by following their development in a more or less natural but human-controlled environment. Of course, when students work on these farms and not only come as observers, further educational objectives can be achieved.

With the spread of environmental education, some school farms, especially in Western Europe, have left some areas to go "back to nature" or have landscaped various habitats for students to study. Several school farms let students set up *ecological experiments* to investigate the effect of human interference with nature.

Supply gardens

Many school gardens *supply teachers of biology and agriculture* with specimens of plants that would otherwise be difficult to obtain. Thus, some school farms have specialized in growing plants according to national curricula in biology and agriculture, even on a regional basis.

During the two world wars, when *food supplies* were short in Europe, school gardens or farms were used to grow vegetables and other agricultural produce to supply the school kitchen or the homes of students. Similar school farms have also been set up in developing countries where they often help schools to balance their budgets, especially boarding schools. In some cases, school supply farms have helped to show students that agriculture can be a profitable enterprise. In other cases, the management of the school gardens and its agricultural achievements have been weaker than that of farmers in the surrounding areas, and this had a rather negative effect on students.

Aesthetics in the school gardens

Besides their other aims (learning about living organisms, work experience, getting to know agriculture, experiencing the joy of creation, supplying produce, etc.), school gardens can contribute to aesthetic education in two ways: they can be beautifully landscaped and they can produce cut and pot flowers which students can take home. Where this is done, curricula have also been developed with the objective of teaching students how to grow plants at home and how to improve the home environment by planning and planting a garden.

School gardens to assist science education

Simple experiments in school gardens can provide science teachers with a unique opportunity to set up *meaningful experiments*, based on sound experimental procedures in which the results have a high practical significance. Often, the results (e.g. plants ripening at different times after experimental manipulation) can be much more impressive and significant than experiments in the laboratory. They can be used to train students in *decision making*. In the school garden, many biological principles can be practically *applied*, thus helping to *transfer learning*. (The value of these instructional goals is discussed in the sections Enhancing the transfer of learning, p. 22, and Learning to make decisions, p. 29.)

COORDINATING AND INTEGRATING SUBJECTS IN THE AGRICULTURAL CURRICULUM

Subjects are said to be *coordinated* with each other when their curricula are planned together but taught as separate subjects. *Integrated* curricula include subject matter from other subjects which are closely related to the main subject and are taught together. In some schools, agriculture and science have been integrated into one subject – agricultural science. In others, they are coordinated. If one or the other is not done, something is wrong with the curriculum.

While agriculture can contribute to *science education*, science is absolutely essential for understanding modern agricultural practices and applying them to the benefit of farmers. Science education thus has an important position in the agricultural school curriculum. Before discussing how agriculture and the sciences can best be coordinated, it is necessary to consider the differences between them.

Agriculture as technology

Technology is usually defined as an *application of science*. In reality, technology is older than science. Wise people knew how to heal before medical academies were founded. Parents and teachers educated children before pedagogy was recognized as a science.

The same is true for agriculture. The "invention" of sowing was one of the greatest human achievements. Astute farmers knew how to select the best seeds for propagation long before genetics was developed, and they made cheese and many fermented products long before biotechnology became a scientific discipline.

What, then, is the main contribution of modern science to agriculture as a technology which preceded it? It is scientific methodology. Before the scientific revolution, progress was based on *experience* alone, which is an

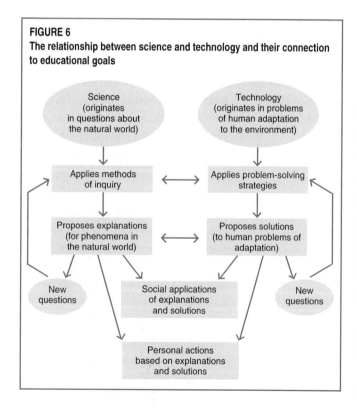

FIGURE 6
The relationship between science and technology and their connection to educational goals

accumulative and, therefore, time-consuming process. Modern science is based on *experimentation.* Both experience and experiment have the same semantic and substantive root. Both demand careful observation and an intelligent drawing of conclusions.

In spite of these similarities, there is a fundamental difference between science and technology. The purpose of science is to generate knowledge and to organize it according to generalizations which enable a further generation of knowledge. The purpose of a technology, on the other hand, is to solve problems. Therefore, their methodologies must be different: inquiry in the case of science and problem-solving strategies for technology. Science produces explanations of phenomena; technology develops solutions to problems. However, in both cases the end product is a new cycle of investigation. Every new scientific explanation arouses new questions and, when technological problems have been solved at a certain stage, new problems arise at a later stage and have to be solved in their turn.

In the sciences, we ask out of curiosity: *how can we explain* phenomena? In agriculture, we ask three practical questions:

- *How can we change* something in the natural situation to serve our needs and goals?
- Is the possible change *also worth while* from an economic point of view?
- *Should we proceed* with the change, taking into consideration the protection of the environment for the future?

Coordination between science and agriculture

Probably the most crucial issue in agricultural curriculum development is the relationship between the sciences, mainly biology, and the agricultural subjects. In many agricultural schools, the three-tiered system of agricultural universities is copied: first, the "propaedeutic" science subjects such as physics, chemistry, biology and sometimes soil science are taught. Only after that are the basics of plant production and animal husbandry introduced; and, in the third step, specific crop production and "auxiliary subjects" such as economics, accounting and extension are added.

Trying to arrange the *sequence of subjects* in the curriculum in this way poses several problems. Usually the time available (two years for a certificate course, three years for a diploma course) is not enough to build such tiers. Furthermore, studying one of the basic science subjects takes more than a year, especially when several have to be taught at the same time (e.g. chemistry and biology). Chemistry is particularly problematic when it is taught in the "classical" or "logical" way: first general and inorganic chemistry, then organic chemistry and, finally, biochemistry. Often the teacher does not reach the last part, which is most important for agriculture. This is not necessarily the only way to teach chemistry. When teaching about acidity in general chemistry, for instance, it is possible to introduce the example of organic acetic acid (vinegar), with which students are familiar.

Much of what is taught in a regular secondary school or college chemistry course is irrelevant for agricultural schools. On the other hand, parts of chemistry which are very important for agriculture, for example soil chemistry, are not included in most regular chemistry courses.

It is simply impossible in an agricultural school to teach a "full" science course which would constitute the curriculum for students who major in the subject at a university. Rather than "water down" such a specialized course, the science curriculum in the agricultural schools should concentrate on those chapters which are most important for agriculture, while also taking into account emerging topics such as biotechnology.

The argument that agricultural schools should offer a course similar to that taught in a college so as to enable the best students to continue their studies at a higher level does not warrant an irrelevant course for the vast majority who will not have that chance. Those who will go on (something to be encouraged!) can be prepared for upgrading without having the majority of the class study a programme which does not focus on the main goal: to train agriculturists at the level the school is supposed to offer.

When biology, soil science and plant production are taught separately by different teachers, duplications often occur. Where should mineral nutrition be taught? In soil science? In chemistry? In physiology? In basics of plant production? When should the needs of specific crops be taught? Obviously, the topic will reoccur at different stages of a *spiral curriculum* (see the section Basic principles of curriculum development, p. 39). At each higher level, there is a chance to reinforce earlier learning and to apply previous knowledge. However, to do this, both the different course curricula and their teachers must be well *coordinated*. It often happens that one teacher is not aware of what another has taught at an earlier point. In such a case, duplicity (and a waste of time) occurs more often than reinforcement.

The problem is worse when *teachers use different terms* for the same concept. For instance, the chemistry teacher speaks about cations and anions, but the physiology teacher uses the terms positive and negative ions. Bright students might ask if these "two pairs of ions" have something in common, and the confusion could be cleared up. However, when no one asks and students learn by rote, they can easily make incorrect conceptualizations.

The solution to these problems can be found mainly in the *integration* of close subjects and in *teacher cooperation* in curriculum development and implementation.

A major problem that restricts the chances for an integrated approach lies in the internal structure of the school. When the school has a department of science that is separate from the crops and animal husbandry departments and perhaps even more departments, there are not many opportunities for teachers to meet and discuss curriculum issues across the board. *All* science and agriculture teachers should meet together and look first at what students must learn in agriculture to become good crop and livestock producers and then decide what scientific prerequisites are needed.

Biology can be studied at different levels of organization of plants and animals – from the molecular and cellular to the ecosystem level. Much of biology at the level of organisms and their environment can best be taught in connection with plant or animal production, respectively. Thus, the biology teacher can concentrate more on cellular biology, with a special emphasis on the repercussions for applied genetics – and especially breeding – and on the ecosystem level, with implications for environmental awareness.

Agricultural chemistry courses should be planned jointly with teachers whose students' will later need to know some chemistry, for example to understand soil-plant relationships, and plant and animal physiology.

Much of *plant and animal physiology* is best taught integrated in a *basic crop or animal production course*, respectively. Specialized courses, for example horticulture, irrigation (where appropriate), plant propagation, plant protection and crop storage can then be built on the basic crop production course, while the course on animal production and husbandry can serve as a basis for specialized studies on animal health, pasture management and the like.

Of course, the curricula in the *applied social science fields*, mainly farm management and extension, should also be integrated with relevant basic sciences courses: for instance, basic economics and accounting, for farm management; and elements of rural sociology, social psychology, communication and adult education that are needed for agricultural extension. At the same time, farm management must also be coordinated with plant and animal production. *Integrated courses* take up more hours in the weekly timetable of the school. This enables teachers to know their students better than when they see them just once or twice a week. It also gives more opportunity to apply scientific principles in a large array of relevant situations. However, integrated courses require a wider scope of knowledge.

A number of other subjects besides science have an important place in the curriculum of agricultural schools but, in the spirit of increased integration, it is better to connect specific items of these subjects to the closest topic in one of the agricultural courses.

Environmental education

Environmental education is an important topic which must be a part of any *sustainable development* course. For this reason, it should not be taught as a separate subject. Rather, environmental issues should be discussed whenever an opportunity arises during science or agricultural classes. Environmental education is concerned, above all, with creating an *awareness* of the vulnerability of nature and the need to preserve it for future generations. Of course, awareness is not enough. Students must learn how to *identify* environmental problems, how to *analyse* them in order to find an *optimal solution* that considers the interests of groups of people who might be affected.

One of the most important tasks of an agricultural school is to teach how to *balance agricultural production with the protection of natural resources*. We can no longer advocate maximum production either from an economical or an ecological point of view. Maximum production has led to unsellable surpluses in the resource-rich countries that could afford the high input prices. Most of the farmers in developing countries cannot pay the high prices for inputs needed to attain a maximum production level where the returns are marginal. Also "maximum protection of natural resources", meaning the total conservation of nature without allowing any human activity, is not feasible. Therefore, the solution can be found only in an *optimized system* of production and protection of natural resources or, in short, sustainable agriculture.

Nutrition education

Agriculture produces mainly food for subsistence and for the market. Agricultural extension workers, who are trained at an agricultural school, are often called on to assist not only farmers in the field but also farming families. Improvements in living standards often begin with improvements in the level of nutrition. A family that suffers from malnutrition (not necessarily undernutrition!) cannot exploit its agricultural potential. Of course, the agricultural school cannot convert every agricultural student into a nutritionist. However, future extension workers, agricultural teachers and administrators should be aware of the food and nutrition problems facing the rural community with which they work and should appreciate the role of agriculture in their solution. For this purpose, they must understand the nutritional value of foods and the functions of the different constituent nutrients so that they can advise on the formulation of a balanced diet for different groups and take nutritional needs into account when assisting in home garden planning.

In many cases the "acting farmer" is the same as the "nutritional expert" in the family: the woman of the house. This is an excellent opportunity to discuss *gender issues* in the agricultural school. Agriculture is not a male prerogative and nutrition is not an exclusively female domain. Therefore, the addition of a home economist or nutritionist to an agricultural curriculum development team is a great asset.

Many relevant topics of nutrition can be discussed in relation to plant and animal physiology, for example the protein content of different crops and varieties, the nutrient requirements and the digestion of food in mammals during different stages of their life. Students' interest is usually enhanced when different farm animals discussed in class are compared to the human body. Crop protection should include postharvest storage of food to protect against insects, rodents and moulds and their hazards to consumers.

However, nutrition education in the agricultural context is not only a biological issue. When teaching agricultural economics, the influence of market prices on supply and demand and the advantages of successful on-farm storage of food for the family should be discussed.

Health and population education
Nutrition education can be seen as a part of health education, but the latter goes much further. When teaching about

animal diseases and their causes (mainly bacteria, viruses, worms and, of course, nutritional deficiencies), parallel problems in human medicine should be discussed. Therefore, in various agricultural curricula, animal and human biology and hygiene are treated together.

In many countries, improved health services have contributed to an increased population growth. As much as the former is a blessing, the latter is dangerous, especially when the production of food and the chances for employment do not grow at the same speed. Therefore, agricultural education cannot ignore the population issue.

Balancing bio-based and socio-economic topics in the agricultural curriculum

Agriculture as a technology is not only based on biology and other natural sciences. In practically all agricultural situations in which a decision has to be taken, socio-economic considerations play an important role. This is even the case when the issue seems to be biological-ecological, for example in plant protection, the feeding of farm animals and soil conservation practices. In all these cases the question of cost and benefit has to be considered, as does the social price to be paid and even the cultural implications. It is therefore necessary to consider carefully which topics from the social sciences need to be taught and where they are best introduced into the curriculum. Teachers who were trained mainly in the natural sciences, with little exposure to economics and other social sciences, often hesitate to integrate socio-economic aspects into their agricultural curriculum. The best place to bring the two facets together is in farm management. However, it is not enough to restrict the study of socio-economic issues to a specialized course such as farm management because these issues have to be faced in all kinds of situations.

Integration with field activities

After having discussed the need and ways for integration between agriculture and related subjects, we must consider an integration problem which is usually much more complicated – coordination between classroom teaching and instruction in the field. The big advantage of an agricultural school with a school farm is at least partly lost when the theoretical classroom curriculum is not coordinated with field activities. The main problem often lies in the higher status of the agricultural teacher compared with that of farm instructors. In such cases, instructors sometimes feel that teachers contradict their view of the farm situation, *if and when* teachers take their students out into the field. What students see in the field is not always congruent with what they are taught in class. This is another reason why instructors and teachers should have a common forum to discuss the field situation and why a close working relationship should exist between teachers and farm instructors. (The issue of teaching in the field is discussed in more detail in the section Teaching on the school farm, p. 97.)

Team teaching

Team teaching represents an effort to capitalize on the special talents, interests and experiences of individual teachers and thus to improve their effectiveness. Team teaching is especially important for the induction of new teachers. Not all teachers of agriculture will have enjoyed full pre-service pedagogic training. In some countries, former extension workers or other employees of the Ministry of Agriculture are assigned the post of agricultural teacher without a systematic teacher's training. Learning on the job in cooperation with a team of experienced teachers can work wonders in such cases.

Team teaching contributes to the professional growth of all teachers involved, enabling them to receive reciprocal feedback. At the same time, students meet more than one teacher in a given subject field, so each has a better opportunity to relate to a teacher whom he or she finds more stimulating.

Team teaching does not necessarily mean that two teachers are always working with the same class. That would be too expensive. Several degrees of team teaching have proved successful.

Team planning. Agricultural and science teachers, sometimes supported by invited curriculum and subject matter specialists, jointly plan the curriculum and sometimes also their lessons, including the preparation of audiovisuals, the ordering of laboratory equipment and the planning of joint excursions. However, the actual teaching will be carried out by a single teacher.

Sharing class teaching. Two or more teachers will join in planning the lessons but each teacher will teach those topics that interest him or her most or in which he or she is most qualified. In this way, each teacher will become familiar with the entire programme but will concentrate on his or her special area of expertise. This arrangement may cause difficulties in planning the timetable and it requires a high degree of coordination among the cooperating teachers. However, it cuts down the number of subjects an individual teacher must master. In such a sharing arrangement, it is important that teachers make sure they use the same terminology for the concepts they are teaching.

Concurrent teaching. Two teachers, for instance a biology teacher who is devoted to the idea of sustainable development and a social science teacher who is concerned about the need to solve the problem of landless farmers, discuss a plan to clear the bush in an area which is prone to soil erosion. Both relate to the project from their divergent points of view. This method can yield a lively classroom discussion and demonstrates how people can begin from different starting points and reach a compromise. In most schools, this kind of team teaching is possible only in a number of lessons where the topic warrants such a confrontation of views.

Sorry, no blueprint

There can be no blueprint for curriculum development in an agricultural school. The scope of subjects, the measure of integration among subjects and between the classroom and the field all depend on the philosophy of the school, the level of education, the composition of the students, the background of teachers and their willingness to participate in curriculum development, and the opportunity given to them by the principal. These and many more factors influence school-based curriculum development, even if it only concentrates on the adaptation of a national curriculum to the specific agro-ecological and socio-cultural situation of the school. It is best for each school to specialize in topics in which it has a relative advantage over other schools, either because of its geographical position or because of human factors.

Curriculum development is an art in itself. A guide such as this cannot go into all the detailed procedures of good curriculum planning. Asking the assistance of a curriculum expert can be useful, but more important is the readiness of teachers to work as a team on the school curriculum. A group of teachers who devised such a school-based curriculum in an African agricultural school summarized their experience by writing: "The people who developed the curriculum were not experts in curriculum development .... It is believed that there are quite a number of aspects which will need improving ... (but) the process of developing this curriculum was *a valuable staff training experience* in itself."

LESSON PLANNING – THE FINAL STEP IN CURRICULUM DEVELOPMENT

Lesson planning is the last part of curriculum development. It is the fine tuning, without which the best curriculum guideline remains blurred. It is the crucial step from curriculum development to curriculum implementation in the form of teaching.

Lessons are planned for a particular time frame. Teachers usually plan their teaching year by year or term by term, taking into account external time barriers such as examinations and holiday periods. This overall plan is then broken down into smaller units – topics and daily lessons. Each unit should have its specific objective but also be embedded in the stream which flows from previous lessons to the following ones, thus reinforcing earlier learning and providing an advance organizer for the following units.

Following are the several steps included in lesson planning, although they do not necessarily occur in this order:

- defining the main objective and the impact points of the lesson;
- reviewing the content or subject matter to be taught;
- selecting and gathering materials that can help to make the lesson clear and effective;
- choosing the most suitable teaching style;
- preparing or choosing the most appropriate teaching props;
- planning the division of time.

It is important to choose a few *impact points* and to make sure that these are well understood. Trying to pack too much information into a lesson will miss the target. Making a *timetable* helps teachers to review their preparations and consider whether all the material can be used in one unit or whether enough learning activities have been prepared to cover different groups of learners who might need different approaches. Timetables are also useful during a lesson to check whether a topic is being covered within its allotted time. Timetables can be used again with a new group of students. In this case, lesson planning is above all an exercise of adaptation to circumstances which generally change from one teaching experience to the next.

Lesson plans usually have three parts:

- a short introduction, which should not last more than a few minutes;
- a central part, which is divided into several teaching steps; and
- a concluding part, which should include a self-evaluation by the teacher.

The *introduction* serves several purposes:

- to connect the lesson with the previous one, perhaps by reviewing students' homework or testing what students remember;
- to determine whether students are ready for the new topic, especially if they were asked to prepare themselves for the lesson;
- to tell students what the new topic is about and how it connects to previous learning (setting an advance organizer).

To encourage motivation, the opening of a lesson should be thought-provoking and arouse interest. It is usually a good idea to ask students what they know already about the new topic, thus inviting their active participation.

Each of the teaching steps in the *central part* of the lesson should treat a specific impact point, but they should all lead towards the central objective of the lesson. The teacher should take into account that not all students learn in the same way. Some perceive better by listening, others by viewing. Therefore, the aids and props used should vary. An explanation that is persuasive for some students might not be so for others. Therefore, a good teacher plans for alternative ways of teaching an impact point.

The *concluding part* of the lesson serves several purposes. The teacher recapitulates the different steps and emphasizes how, together, they form a meaningful unit, so students have an idea of the whole. The teacher checks that they have understood the central message and can draw conclusions from it. This is also the point at which teachers set tasks by which they can assess whether students are able to use the intellectual skills they have learned, for instance if they can apply newly learned generalizations. According to this first evaluation and the objective of the next unit, the teacher assigns homework or other preparatory tasks.

Taking notes

Should students *take notes* or receive a written summary at the end of the lesson? Both methods have advantages and disadvantages. Taking notes is an important learning skill which must be guided and exercised. Agricultural school graduates will need this skill on many occasions after leaving school, for example when attending extension meetings, in-service courses, field days and during rural appraisal work. Taking notes can help students pay attention to the lecturer but it can also have the opposite effect of distracting students from paying attention to the key points of the lecturer and causing them to lose the gist of the lecture as a whole. This happens when students do not know how to take notes and try to write a verbatim protocol instead.

Therefore, it is imperative that teachers check the notes students make during the lesson and, if necessary, suggest better ways to do so. Once students master this skill, their notes can be excellent for reviewing the lesson, especially if no other written summaries exist. Alternatively, teachers can hand out a short, mimeographed or otherwise duplicated outline or summary of the lesson. This has several advantages: it shows students how to differentiate between key or impact points and insignificant parts of the lecture, and how to summarize a lecture. This is especially important when teachers think students do not yet possess this skill. Lesson summaries also help to provide a *closure* to the lesson and to leave students with something solid to review, especially in the absence of a suitable textbook.

Summaries are especially useful for teachers who use a guided inquiry style and do not want to reveal the results of an inquiry before students have attempted to find the solution by themselves. In any case, the summary should be simple. When it is handed out to students, the value of a summary can be augmented by the inclusion of drawings. Of course, teachers can also write the summary on the blackboard for students to copy or they can dictate the summary, but both these options take valuable time from the lesson.

Following the lesson plan

The timetable, with its preplanned steps, helps teachers to be better aware of what is actually happening during the lesson. They may ask themselves: Am I following the plan? Does it make sense to follow the plan? Perhaps I should change it in the light of the students' reactions? Are they following me step by step? Are they active in the way that was planned? Are they motivated, interested? Is anything taking a different direction from that planned? Why? What should I do about it (if any specific action is warranted)?

As much as teachers plan and teach accordingly, this is not necessarily what students learn. On the other hand, students learn things that were not planned. This is sometimes called the *hidden curriculum*. The outcome of mathematics teaching, for instance, is often that students *hate* mathematics. What are their reactions to our teaching of agriculture? It is better to be sure.

Planning a field lesson

Planning a field lesson demands the same preparation as a classroom unit – and more. Outside the classroom, motivation might be heightened but teachers must also be aware that there are also more distractions. Therefore, the need to plan each step carefully is enhanced. Before the lesson it is especially important to check that all the tools needed for a demonstration are ready and that there are no technical obstacles that might inhibit the lesson, for example if the demonstration field has just been irrigated and cannot be used. Where a field instructor is responsible for the school farm, or if the lesson is to be held on a farmer's field, full coordination with the people concerned is absolutely necessary.

Of course, all the rules for classroom and outdoor teaching also apply to work in the *laboratory*.

BIBLIOGRAPHY   **Bloom, B.S., ed.** 1956. *Taxonomy of educational objectives. Handbook I: cognitive domain*. New York, McKay.

Blum, A. 1991. Agriculture: educational programs. *In* A. Lewy, ed. *The international encyclopedia of curricula*, p. 926-927. Oxford, UK, Pergamon.

FAO. 1982. *Food, nutrition and agriculture. Guidelines for agricultural training curricula in Africa.* Rome.

FAO. 1992. *Planning for effective training – a guide to curriculum development.* Rome.

FAO/USAID/University of the Philippines. 1981. *Food, nutrition and agriculture. Guidelines for curriculum content for agricultural training in Southeast Asia.* Rome, FAO.

Lewis, R. 1984. *How to help learners assess their progress: writing objectives, self-assessment questions and activities.* London, Council for Educational Technology.

Popham, W.J. 1975. *Educational evaluation.* Englewood Cliffs, N.J., USA, Prentice-Hall.

Race, P. 1992. *53 interesting ways to write open learning materials.* Bristol, UK, Technical and Educational Services.

Raman, K.V. & Sudarsaman, R. 1990. *Curriculum development for higher education in agriculture: arising issues.* Rajendranagar, Hyderabad, India, National Academy of Agricultural Research Management.

Smith, S. & Taylor, B. 1991. Curriculum planning for women and agricultural households: the case of Cameroon. *In* C.H. Gladwin, ed. *Structural adjustment and African women farmers*, p. 373-386. Gainesville, USA, University Press of Florida.

Tomlinson, P. & Quinton, M., eds. 1986. *Values across the curriculum.* London, Falmer.

Tyler, R. 1949. *Basic principles of curriculum and instruction.* Chicago, USA, University of Chicago Press.

Wallace, I. 1992. Agricultural education as a learning system in Africa: enhancing effectiveness through innovations at the formal/non-formal interface. *Int. J. Educ. Dev.*, 12: 51-64.

Warwick, D. 1987. *The modular curriculum.* Oxford, UK, Blackwell.

Teaching styles and skills

**TEACHING –
SCIENCE, ART
OR CRAFT?**

Some people consider teaching to be a science and emphasize the usefulness of learning and teaching theories and the contribution of scientific research to the improvement of teaching. Others believe teaching to be an art because it involves human beings, their feelings, emotions and values. Teachers have much scope for intuition, improvisation and creativity, which are commonly considered to be ingredients of art.

Teachers have often been compared with medical practitioners and engineers, both of whom use scientific principles but operate with dexterity, resourcefulness and intuition. Although teaching lacks the highly developed scientific bases underlying medicine and engineering, it has similar or even more artistic elements. It could also be compared with agriculture as a profession where many scientific principles are used but where one also needs "green fingers". Since teaching falls somewhere between a science and an art, one might look at teachers as highly skilled artisans – a term which is not far from that of artist – who create something useful according to prior specifications, yet with a personal, individualistic and artistic touch.

Teachers differ in their teaching styles and in the teaching skills they possess. A *teaching style* expresses a teacher's *basic approach to teaching*, which may be general and constant or vary according to a specific educational goal. For instance, some teachers might always use an authoritative style because they believe this to work best in their situation or culture, or they may use this style for certain topics and situations but then change to an interactive or even discovery teaching style when they think these are more appropriate.

Teaching skills express themselves mainly in the *things a teacher does* while teaching. These can be observable techniques such as asking questions, giving cues, etc. and the use of teaching aids (which is treated in a separate chapter). Other skills, for example analysing a teaching-learning situation or building a conceptual framework, cannot be observed and are therefore more difficult to teach. Creative or imaginative skills are traits with which a gifted teacher is "born". Training can only help to develop these traits further.

TEACHING STYLES

Authoritative versus interactive teaching

Authoritative teachers are aware of the huge difference between their knowledge and that of students. They know what *they* want to achieve and believe that the best and quickest way to teach is to present the knowledge in a straightforward lecture or to let students read a chapter in a book. They assume that, if students listen or read carefully, they will understand the content of the lecture or the book. If students do not understand, it must be because they lack the necessary intellectual ability and probably should not be in their class. They do not like students to interrupt the flow of their lecture with questions, these can come at the end of the lecture. Rather, students should take notes and then think over by themselves the points they did not understand.

This description of authoritative teachers might be extreme but there are many cultures in which this is the accepted style. If they do not understand, students consider it to be their own problem. Rather than losing face by asking for clarification (especially in public), they make an effort to catch up with those who have a quicker comprehension. They certainly are not expected to question the truth and validity of the teacher's expositions.

In many traditional Western institutes of higher education, lecturers lecture and professors profess their knowledge, often from a raised platform to symbolize their superior standing *vis-à-vis* the mass of students. In spite of this, great scholars have come out of authoritative teaching systems. They were probably the superior students in a system where the best and not the majority count.

Interactive teachers are also in control of the classroom, but as educational leaders. They believe that what students (and hopefully as many as possible) *learn,* and not what they themselves know and teach, is important. To learn, one must be intellectually active, and interactive teaching encourages students to become active and even critical learners.

Although the two approaches differ in their basic philosophy, they do not exist only in their extremes. Authoritative teachers may use demonstrations and props to illustrate their expositions. On the other hand, interactive teachers may employ the "pure" lecture method on appropriate occasions. To hear a gifted orator sum up an area of knowledge in a well-prepared and provocative lecture can be an experience to remember throughout life.

However, if authoritative lectures are not given by an outstanding speaker and are not accompanied by impressive and illustrative teaching aids, they could be replaced with tape-recordings which students can listen to at their own convenience. The advantage of the tape-recorded lecture is that difficult parts can be replayed. Interactive teachers, on the other hand, will involve students by posing questions of different types (see the beginning of this section). They are prepared to build on students' responses and to interpret them. Interactive teachers must be quick to pick up not only the meaning of students' questions and responses, but also *why* students ask and respond in a particular way – and they must be prepared to change the preplanned course of the lecture, if necessary. However, teachers must also be careful that one or two students do not "take over" and disturb all the others.

Interactive teachers create a more democratic atmosphere. If the class is small, they will prefer to seat their students in a horseshoe arrangement in order to have better eye contact with them and to avoid the "teacher opposite students" situation of a regular lecture room.

In the *Socratic style,* teachers involve students actively in a dialogue, but they do so to lead them to what the teacher believes is absolutely true. The teacher makes a point of revealing students' hidden ignorance and "wrong thinking" in order to bring them to the truth which should have been obvious to them, had they thought rationally. In this style, students have no opportunity to think creatively.

Guided inquiry or discovery teaching

Inquiry or discovery teaching goes a step further towards students' active involvement. Inquiry teachers neither lecture dogmatically nor lead their students in a Socratic fashion towards a preset conclusion. Rather, they let their students discover phenomena and generalizations by themselves. These are assumed to be "hidden" somewhere, and students have to discover them through methods of scientific inquiry. In doing so, students are guided by the teacher at various levels.

Three main reasons speak for inquiry or discovery teaching:

1. The revisionary character of science. Students come to school with the assumption that scientific knowledge is absolute and reliable. Often this misconception is reinforced in school, where there are either "right" or "wrong" answers. In reality, knowledge, and especially scientific knowledge, is correct only until new insights, based on new inquiries and new explanations, are obtained. In earlier centuries, it took a long time for accepted knowledge to be proved wrong and new discoveries accepted. In our scientific age, a flood of new scientific findings forces us to revise our knowledge constantly. This is especially true for agricultural knowledge which is becoming more and more scientifically based. It is important for students to understand the temporary character of specific knowledge. They must understand that discovery does not happen in one day; discovery comes after a process of inquiry, which begins with doubts of the validity of present knowledge and continues with hypothesizing, experimenting and weighing up alternative explanations of the results obtained in an experiment. It is a thorny way through doubt and failure, until a new, again temporary, "truth" is found.

2. Acquiring inquiry skills. As graduates of agricultural schools, students will work in different positions which force them to be at the forefront of agricultural knowledge. In these positions, they will not only have to understand how this knowledge was obtained, they will also be involved in generating solutions which are not "in the book". They should be able to set up simple field tests and, to do so, they must understand the rules of scientific hypothesizing, experimenting and evaluating research results. These skills can be acquired only through inquiry training.

3. Discovering is fun. Discovery learning is not only a challenging intellectual experience and an opportunity to acquire a skill that is crucial in generating or assimilating new knowledge; it is also great fun for students. Never mind if a clever researcher has discovered a certain fact or generalization before. For the student who rediscovers the same thing, but by his own efforts, the joy is the same.

For practical reasons, inquiry teaching may be divided into three main stages:

- discussing the problem, phrasing a hypothesis and planning an experimental design;
- setting up the experiment and performing all the work connected with it until the findings are recorded;
- analysing the data and drawing conclusions.

Of these three steps, the first and third can be carried out in the classroom, but the performance of an experiment in agriculture requires a laboratory or a "land laboratory" – an experimental plot on the school farm. Experience has shown that agricultural field trials are often more impressive than laboratory experiments. They are closer to the central focus of the school and of its students, i.e. agriculture. They also usually have more practical meaning than laboratory experiments.

The main problem with inquiry teaching is the time and the equipment needed for the execution of experiments. Because of this, the experimental stage may sometimes have to be replaced by supplying students with the results they would have obtained if they had actually conducted the experiment in the laboratory or in the field. This is done in the form of *narratives of inquiry* or *invitations to inquiry*. These teaching methods are described in the section Textbooks and other written sources, p. 109. However, some inquiries should be conducted in full. *Field trials,* as described in the section Teaching on the school farm, p. 97, are the best way to do this in an agricultural school.

Inquiry guidance can be at three levels:

- The teacher poses the problem and describes how to go about solving it. Students have only to find out the answer.
- The teacher poses the problem, but students have to

work out the method to solve it, as well as finding the answer.

• All three steps are left open for students to discover.

Cultural congruence – a problematic issue in the choice of teaching style

Teachers often come from an environment which is socio-culturally different from that of their students. In such a case, teachers and students are said to be *culturally incongruent*. This is more common in developing countries, where socio-cultural differences between social strata are often more pronounced than in the industrialized countries. Teachers coming from a middle-class background may be unaware of the traditional way of thinking which their students bring with them from home. Often the values that prevail among these teachers run counter to those held by the students' families. When teachers use a teaching style that runs counter to what their students experience in day-to-day activities, the effectiveness of teaching is often diminished.

This can be a problem when students encounter inquiry learning for the first time. However, when it is introduced in a challenging way, students tend to take it up readily because, in all cultures, there are people who like to try out something new. A good point of entry for introducing inquiry learning is a scientific achievement that has helped to advance agriculture and that addresses an agricultural problem to which students can relate.

In some traditional societies and environments, children tend to be very sensitive and to prefer personal and informal relationships with authoritative figures. These children are motivated by personal and family experiences as opposed to the unemotional presentation of information they might find in schools. In other cultures, students are already trained at home to accept whatever parents and teachers say, without daring to ask for justifications or explanations.

The problem might be greater in elementary schools, where the cultural shock is stronger. However, especially with agricultural school students who come from traditional, rural families and have not been "conditioned" by Western styles of teaching during their earlier school years, cultural incongruence might still be a problem.

When teachers are aware of the disparity between their cultural values and those of their students, they should assess both their students' and their own cognitive styles and think of ways to bridge the gap.

TEACHING SKILLS

Teaching skills are techniques and practices which a teacher uses to facilitate students' learning. The term "teaching skills" suggests that they can be taught until a certain level of proficiency has been reached. This is mainly true for those techniques of behaviour which can be observed. The use of imaginative practices by teachers mainly depends on their intuition and creativity.

Teaching techniques

The reader will observe that a number of teaching techniques touch on issues that have already been discussed at the theoretical level as principles of teaching. Others have to do with the use of teaching aids, classroom management and evaluation techniques, which are treated in separate chapters. Here, we look at behavioural techniques which are used by teachers when standing before a class and which can be taught, mainly in microteaching exercises, and can be observed by peers or coaches. The categories into which the techniques are placed (see Box) are not mutually exclusive.

1. *Motivational skills.* Varying the stimuli, e.g. variations in movements, gestures, use of stimulating materials that appeal to different senses; reinforcing student behaviour; accepting and supporting feelings that students express; showing enthusiasm and empathy with students; encouraging students' involvement and initiative.

2. *Set induction.* Introducing a lesson by clarifying its objectives and using advance organizers; relating the topic of the lesson to previous knowledge and skills.

3. *Use of communication skills.* Dramatizing and explaining with the help of examples and teaching aids.

4. *Small group and individual instruction.* Organizing and handling small groups and individual students for independent learning and counselling; encouraging cooperative activity.

5. *Closure.* Helping students to link old and new knowledge by reviewing and applying existing understanding to new situations or cases.

6. *Non-verbal cues.* Reducing the amount of talking by the effective use of proper pauses, body movements, facial expressions and gestures. Unlike the techniques listed in No. 1, here the gestures and cues stimulate students to become active and give meaning to the teacher's non-verbal behaviour.

7. *Encouragement through praise.* Praising and accepting students' willingness to respond to questions and to participate on their own initiative, but also giving non-verbal cues to express praise and encouragement, e.g. nodding and smiling to reinforce students' readiness to become active learners.

8. *Fluency in asking questions.* Frequently asking questions which attentive students can readily answer, but without unnecessary repetitions.

(continued)

9. *Higher-order questions.* Asking questions that require students to use higher mental abilities such as application, analysis, synthesis, critical thinking, problem-solving and decision-making skills.
10. *Divergent questions.* Posing questions to elicit students' responses that are imaginative, creative, unconventional and cannot be judged simply to be "right" or "wrong".

Intellectual and creative teaching skills

Intellectual teaching skills, such as identifying and rectifying students' misconceptions or analysing a learning-teaching situation and drawing conclusions from it, cannot be observed and are difficult to develop through training. This is even more the case for the imaginative and creative domain which so often distinguishes gifted or "born" teachers from their average "trained" colleagues.

BIBLIOGRAPHY

Bligh, D. 1986. *Teach thinking by discussion.* Guildford, UK, Society for Research in Higher Education.

Bosworth, D.P. 1991. *Open learning.* London, Cassell.

Duke, D.L. 1990. *Teaching – an introduction.* New York, McGraw-Hill.

Freire, P. 1976. *Education, the practice of freedom.* London, Writers and Readers.

Gage, N.L. 1978. *The scientific basis for the art of teaching.* New York, Teachers College Press.

Habeshaw, T. 1989. *53 interesting ways to help your students to study*, 2nd ed. Bristol, UK, Technical and Educational Services.

Lindsey, C.W. 1988. *Teaching students to teach themselves.* New York, Nichols and London, Kogan Page.

Lorac, C. & Weiss, M. 1981. *Communication and social skills.* Oxford, UK, Pergamon.

Paisey, A. 1983. *The effective teacher.* London, Ward Lock Educational.

Perrott, E. 1985. *Effective teaching: a practical guide to improving your teaching.* Harlow, UK, Longman.

Schwab, J.J. 1964. *The teaching of science as inquiry.* Cambridge, Mass., USA, Harvard University Press.

Wilkins, E. 1975. *Education in practice – a handbook for teachers.* London, Evans.

Teaching
methods

All teaching methods have one common goal: to facilitate students' learning. Since not all students learn in the same way, and some methods are better suited for certain types of learning activities, it is obvious that good teachers must choose for each lesson the methods which are best suited to enhance the chances that effective and efficient learning occurs.

Major teaching methods

- Lecturing
- Using resource persons
- Working with small groups
- Self-instruction and distance learning
- Teaching in the laboratory
- Teaching in the field
- Using the community

LECTURING

A lecture usually consists in one person talking to many about a given topic. Lecturing is probably the oldest method teachers use to work with a group of students. It is an adaptation of our most common means of communication – the spoken word, accompanied by gestures and teaching aids.

Lectures are used in all cultures to communicate accrued knowledge from one generation to another. It costs less than other teaching methods and, especially where textbooks are scarce or available only in a foreign language, lectures tend to become the major source of information for students in agricultural classes.

Teachers often use the lecture method to "cover" as much of a topic as possible. In doing so, they prevent students from learning to "*dis*cover" and analyse problems by themselves – a very important skill for future agricultural leaders (see the section Teaching styles, p. 74).

Lectures may be used effectively and economically to convey information, provide explanations and stimulate interest. However, much depends on *how* a teacher lectures. A "pure" lecture is unidirectional, with no effective feedback. Students tend to remain passive, lose interest and let their thoughts wander, especially if the lecturer talks on without giving them an opportunity to ask questions, clarify difficult messages or participate in a discussion.

The most important essentials in a lecture are clarity and interest, which begin with the teacher making sure that what is said can be heard and what is shown can be seen. The effect of well-prepared lectures is lost if the acoustics are bad or if illustrations shown are too small or blurred. Clarity of presentation also involves explicit language, a clear structure, repetition of the main points, paraphrasing and an optimal use of the voice.

Stimulating interest is linked with the enthusiasm shown by the lecturer. This is expressed in gestures, changes in the vocal tone, eye contact and other non-verbal messages. Suitable examples and teaching aids that match the content of the lecture can increase students' interest. Even so, attention fluctuates after 15 to 20 minutes. This decline in attention is less likely to occur if the lecture includes some short activities for students, for example brief small group discussions – often called "buzz sessions".

Experienced teachers let their students ask questions or add their own ideas whenever a good opportunity arises during the lecture. Others are afraid that the interruption might harm the flow of the lecture and they prefer students to save their questions until the end of the presentation. A middle way is to stop at certain points in the lecture and ask for questions and comments before continuing on to the next point. Above all, interactive teachers make sure that their students not only learn the subject matter taught but that they acquire skills in listening and clear thinking.

USING RESOURCE PEOPLE

The teacher is not the only knowledgeable person who can convey useful information to students. Actually, he or she is often only a secondary source. Especially in agriculture, there are important resource people who can be asked to come to the school and talk to students. They include extension workers, veterinarians, foresters, other experts and community leaders, bank managers, health workers and farmers who have shown initiative and developed inventions.

When inviting these resource people, it is important that the teacher explain to them beforehand the purpose of the meeting with the students and what exactly is expected to come out of it. When a resource person teaches, more time should be planned for questions than in a regular class session.

CASE-STUDIES

Case-studies contain materials in different forms (reports, maps, tables, photographs and so on) which describe a certain case in which important decisions have had to be taken, usually in order to solve a major problem. Case-studies are used to train students in the analysis of a complex situation, with the help of information documents, and in making decisions concerning a real problem. Case-studies are usually based on a specific situation, but they may have to be simplified according to students' ability. The time available to students is also a major consideration when preparing a case-study. Therefore, the amount of materials must be restricted to what is essential to allow students to extract the necessary information.

Case-studies are most suitable for group work. They

give students an opportunity to interact in cases similar to real life situations in which several people are involved in analyses and problem solving. Often, simulation games are based on case-studies, to which they add the advantages of role playing (see the section Exercise games, role play and simulation games, p. 131). Case-studies can also be used as assignments for individuals, especially more advanced students.

WORKING WITH SMALL GROUPS

Working with small groups is probably the setting in which active learning is most facilitated. A typical size for a working group is four to six students. Too many groups in a class make it difficult to supervise and, if necessary, provide input as the groups work their way through a problem. In large groups, there is a tendency for weaker members to remain passive.

The learning achieved in groups depends to a large degree on their composition. In a school situation, *heterogeneous* groups usually work better than *homogeneous* groups. In many cases, one of the intellectually stronger students will take the leadership while weaker students can often learn better in a peer group than in a regular class setting. It is easier for them to show where they have failed to understand a point, and peers who "speak the same language" can often solve the difficulty better than a teacher who might have a communication problem. If the differences in the level of students in a group are not too big, heterogeneous groups perform well. They do not hinder the better students, who often reach a deeper understanding when they explain an unclear point to their weaker classmates.

Group work is especially suitable for problem-solving tasks. Many of the problems confronting agriculture are complex and need expertise in quite different areas – technical, biological, economical, environmental and so on. Therefore, the agricultural extension workers and administrators of tomorrow must be able to work in teams with other specialists. To do this efficiently, they should be used to team work and its dynamics of examining different alternatives and then coming to a balanced decision. Group work in the agricultural school is an excellent way to train students for this type of challenge.

When the advantage of group work is sought for the active acquisition of comprehensive information, the *group puzzle* method can be applied. The teacher prepares background material and sets tasks. The class is usually divided into groups of four to five students and each group is assigned a subtheme. Each group studies its material and its members become "experts" on their subtheme. Therefore, the group is called an *expert group*. In the second part of the exercise, information is exchanged in puzzle groups where one representative of each expert group meets with experts from other groups (see Fig. 7, p. 92). Thus, there is a member of each expert group in each *puzzle group*.

The main effect of group puzzles can be found in the puzzle groups where each member informs the others about what he or she has learned with his or her peers in the expert group, based on the source material. Thus, each student becomes an active peer teacher, acquiring part of the information by studying the sources and then learning the other parts of "the puzzle" from the other students. Students know that the class can only receive all the information when each has acquired an expert status, which he or she has to demonstrate in the puzzle group. This motivates students to work actively and intensively while they are in their respective expert group. The puzzle method is useful when a theme can be divided into subthemes or different aspects.

Seminar work

Seminar work combines a lecture with discussion and group work. It starts with a presentation by the teacher or a student who is well prepared, or with the teacher's review of what students were supposed to study before coming to the seminar. Then questions are posed and a

FIGURE 7

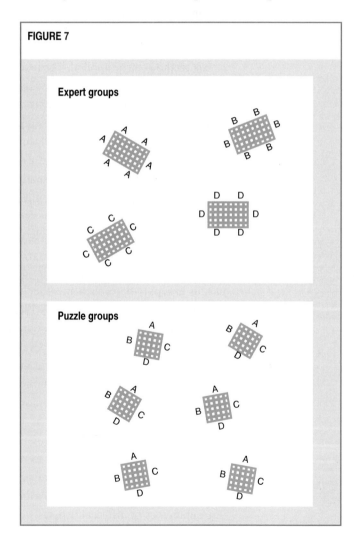

discussion follows. The larger the audience and the greater the dominance of the teacher, the more it resembles a lecture. If the number of participants is small and they are actively engaged in the discussion, the seminar functions as a learning group.

SELF-INSTRUCTION AND DISTANCE EDUCATION

Distance education lends itself particularly to cases in which the number of learners is relatively large, but where they are scattered and cannot leave their jobs for long periods of time. This is more often the case in rural regions than in cities. Therefore, distance learning programmes have been useful in the further education of field personnel, both for brushing up their knowledge and for learning new topics.

Distance learning in agriculture can be used for pre-service training, for example for study towards a specialized certificate, as well as for in-service training. Distance learning demands self-discipline. It is more easily done in a formal framework such as a formal course offered by a specialized distance learning institution. These institutions use a multimedia approach and prepare packages which include written information, audiotapes and, in some cases, even materials for simple "kitchen laboratory" exercises. In many countries, distance learning institutions broadcast radio and television programmes and set up regional centres where distance learners can meet a tutor if they feel this to be necessary. In these centres, students can develop practical skills that require equipment they may not have in their home community.

At the end of each study unit, students complete written assignments which are then corrected by tutors and returned with advisory remarks. In the case of formal courses, grades are assigned, often on the basis of a formal examination. Especially in developing countries, where many students in rural areas do not have an opportunity to reach a higher stage of education, in-service distance education can offer a second chance. It should be encouraged by the educational authorities because those who succeed in distance education show that they have self-discipline, can turn their ambition into positive outcomes and are able to concentrate on a difficult job.

Agricultural schools and colleges can serve as liaison centres for distance learning. Experience with distance learning has shown that some face-to-face contact with tutors enhances the success of distance learning. Distance learners have social as well as purely educational needs. Learners who struggle at home with new materials and unfamiliar problems require encouragement and help. Some of this can be provided in writing. However, meeting a tutor from time to time adds to learners' success and stamina to continue the course to a successful end. While a single agricultural school will not be able to run a full distance learning programme, it can serve as an occasional meeting place between distance learners and tutors.

TEACHING IN THE LABORATORY	There is no need to justify the need for laboratory exercises in the teaching of agriculture as applied science. The problem is rather how to manage them, given the fact that teaching in the laboratory takes more time than classroom teaching and needs equipment that is not always available. The time problem is aggravated when no laboratory assistance is available and teachers have to prepare everything themselves.

The first step towards solving this problem is the teacher's consideration of what purpose laboratory exercises should serve. (Alternatives to laboratory exercises may then be considered, at least for some objectives.) Basically, the laboratory can serve three objectives:

- to *demonstrate* physical phenomena, chemical reactions and biological processes;
- to *train* students *in technical skills*, e.g. microscopy or chemical measurements;
- to let students *discover* scientific facts and generalizations by themselves.

Among the three uses, *the demonstration laboratory* is the most time-efficient. Often, phenomena, reactions and processes can be explained better with the help of a laboratory demonstration. The demonstration laboratory is best used when all the science classes can be held in the laboratory and the teacher can move freely between theoretical explanations and demonstrations.

Sometimes a teacher explains a process or reaction in class and then "proves" its correctness in the next laboratory session. In such a situation, much of the effect is lost. Moreover, not every laboratory demonstration explains what actually happens in a chemical reaction. A demonstration of how the colour of litmus paper changes in a pH measurement according to the acidity of its liquid environment does not elucidate the meaning of pH (the concentration of H-ions in a solution). Nor does it explain the meaning of the pH numbers 1 to 14.

Depending on the knowledge level of the class and the objectives of the course, it may be decided to treat the concept of pH as a "black box". The teacher may be interested mainly in teaching the skill of taking pH measurements of a soil sample and may expect students to know only that the pH numbers indicate degrees of acidity or alkalinity, with 7 as the neutral middle point. In this case, the laboratory exercise becomes part of *skill training* where the main objective is for students to master the technique so they can make their own pH measurements and understand the impact of the results on soil management.

When the main purpose of the laboratory is technical skill training, much repetition is needed. This might be unnecessary in a course for which theoretical knowledge is more important. Thus, in a course for bio-agricultural laboratory assistants, students can be expected to have good working skills in microscopy (e.g. to recognize causes of plant diseases) or in standard chemical tests of milk, and so on. This skill training is too time-consuming for regular agricultural students in a certificate course, who probably will never use these skills in their professional life or who, if they should need them, can acquire them when the need arises.

This point is not always obvious. Some teachers devote a lot of time to teaching students how to use a microscope, although the chances that students will use microscopes in the future are practically nil. Under these circumstances, an explanation of the principles of microscopy, demonstrating the use of the microscope and using enlarged pictures of microscopic photographs might do a quicker, cheaper and no less instructional job. Working with the less "prestigious" but easier-to-handle stereoscope, which gives a three-dimensional and more impressive picture, is often more effective and efficient than spending time on the use of microscopes. Moreover, in most cases there are too few of the latter available.

The importance and advantages of *discovery learning* have been discussed in the section Teaching styles, p. 74. In science teaching and in some parts of agricultural instruction, the best place to let students discover facts and rules by themselves is in the laboratory. (As we shall see, the best place for discovery learning of most agricultural issues is in the field.)

As much as students enjoy the thrill of discovering something which is new to them, and as important as discovery learning is to the development of curiosity and skills in conducting open-ended field tests, the teacher cannot ignore the problem of time, which is much more serious in the case of discovery learning in the laboratory than in the case of laboratory demonstrations. Therefore, methods such as *invitations to inquiry* and *narratives of inquiry* have been developed (see the section Textbooks and other written sources, p. 109) in order to balance the importance of discovery learning with the constraints of scarce laboratory time.

Laboratory work is best organized in groups. This enables better supervision, which is important for safety reasons. It also makes better use of the available facilities. If students are more advanced and can work independently, "stations" can be set up on different benches. At each station a worksheet explains the task, the questions to be answered and the technical steps to be taken.

TEACHING ON THE SCHOOL FARM

Agricultural teachers could well adopt the slogan: "The real classroom is outside. Get out into it!" Most agricultural schools have a school farm and use it for different purposes. In a few countries where agricultural schools deliberately do not run a farm, students join the school after they have already received a thorough apprenticeship training on a master farm. In such cases, the school prefers teachers to take their students to demonstrations given by the very best farmers in the region. These will always be more advanced than a public school farm could be.

The school farm can serve various goals:
• to give practical training in modern agriculture;
• to demonstrate how to grow major and minor crops and rear animals;
• to try out agricultural innovations;
• to create income for the school.

In order to fulfil its educational purposes, the school farm should consist of the following elements:
- a commercial training farm;
- smallholder training units;
- individual students' plots;
- crop demonstration and trial plots.

The status of the farm and the farm instructor

Unfortunately, in many agricultural schools the school farm is considered to be only an appendage and not an integral part of the teaching complex. Thus, the agricultural school often follows the negative image it should destroy, namely that white-collar work is more prestigious than getting one's hands dirty in the field. This attitude is worse when the farm serves mainly as a source of income and students are often considered to be cheap labour. There is nothing wrong with making money on a school farm. On the contrary, how can students learn good management techniques on a farm which is not run economically? The point is that teaching should be no less important than the farm's making a profit.

Much depends on the status of farm instructors or whoever is in charge when students work on the farm. Are they treated as colleagues of the teachers or considered to be "only foremen"? Do the agricultural teachers make a point of inviting farm instructors to join them during visits and requesting their comments on practical issues (in which they may have more experience than the teacher)? These are crucial issues which often demand a new style of cooperation and social outlook.

Practical training

If practical training is part of the school curriculum, it should be treated like other elements in it. Therefore, a detailed list of all the skills a student is expected to learn on the farm must be prepared, preferably with clear indications of what skill levels the student is expected to demonstrate. It is a good idea to *give each student a job list* with these objectives and to ask him or her to make a note whenever these skills are practised. The farm instructor should then indicate at fixed periods the level of mastery (minimal or advanced) reached by the students.

Practical work should be examined and graded and no student should be allowed to graduate without satisfactory grades in practical work. These grades should also take into account factors such as work attitude, cleanliness and efficiency.

Practical training does not mean only working with tools and in the field or with animals; it also includes management training. Students should be introduced to the economic considerations of farm decisions. This again demands a close cooperation between the agricultural economics teacher and the farm manager.

The farm manager should analyse with the more experienced students (usually in the second or third year of their practical training) the actual production costs, including labour and overhead costs, as well as the market opportunities and strategies of the farm.

Agriculture is not only production. Students should learn what is being done on the school farm to conserve and, if possible, improve the quality of the soil. Students should also understand the economic considerations relating to postharvest treatments, including storage and food processing, if they are done on the school farm.

The smallholder training unit

Although students should be trained in modern farm technology, they must also be familiar with low external input methods, as they are practised by many smallholders who cannot afford the maximum technological packages. After leaving school, students might have to farm using simple and intermediate technologies, or they might advise farmers who work under these conditions. They should be aware of the possibilities of farming under suboptimal conditions and become inventive in adapting agricultural technologies to simpler levels.

For this purpose, a small training unit should be established and situated at a site suitable for demonstration purposes and which is close to the school and a road. On this unit, traditional but improved methods and minimal investments should be used. The challenge of this unit is to show students and farmers in the region how traditional methods can be improved, even under the simple conditions which may be typical for the region, for example by using simple irrigation devices or oxen for work. Animal rearing on these smallholder units would be at a similarly low level of external inputs.

Sometimes two such smallholder units are set up on the school farm, both of a size that is typical of the region but each having a different level of external inputs and capital investment.

In this context, it is advisable to develop village technologies (e.g. simple but efficient grain and rainwater storage facilities and local biogas and solar energy units).

The smallholder units should be farmed by students in their first year of school under the supervision of their farm instructor. Students can take turns of one to two weeks in doing all the work on the unit and they should be involved in its decision-making process. Students' work should be assessed by the school farm manager and the head of the crops department. During the rest of the time, students should work on the main farm and have the opportunity to compare the different strategies used, according to the economic and agricultural parameters of the different units.

Individual students' plots

Students should not only work following detailed instructions; they should also have the opportunity to farm a small patch of land on their own and be responsible for their own decisions. They should plan to grow crops on small individual plots, performing all the steps, including the choice and purchase of inputs and the marketing of produce. Work on the individual plot will be done during students' free time, usually in their second year after they have already had practical farming experience on the main farm and the smallholder unit. Students should keep records on all work done on their respective plots and be assessed on the outcome of this work.

Crop demonstration and trial plots

The agricultural school should have a crop demonstration and trial plot to familiarize students with the different crops of the country, especially those which are not grown commercially on the school farm. Changes in the market situation may bring into focus crops which, at the moment, are deemed less important. Therefore, students should be familiar with a wide array of crops and also be able to observe species which are grown mainly for home consumption.

The demonstration area should include trial plots on which new varieties are tested for their suitability to the region. Priority should be given to varieties and techniques which are believed by the experimental stations to be promising but which have not yet been tried out in the school's region. Thus, the school can play an important role in local adaptation studies.

Agricultural teachers should plan these trials with the experiment stations that developed the innovations, with the regional extension services and with their students. The latter should be fully involved in the planning of controlled experiments, including their implementation, in collecting and analysing data and in reporting on the results. Thus, students will have first-hand experience in how to plan and conduct adaptive research and pre-extension trials.

Field days

Farmers should be encouraged to visit the demonstration and trial plots. This can be done in cooperation with extension workers who are active in the region and with farmer leaders. Thus, contact between the school and the farming community and the research establishment can be strengthened, and the school can contribute towards finding optimal solutions to the agricultural problems of the area. Students should be involved in this type of contact as part of their extension and development training. They should listen to what experienced farmers have to say about innovations and the problems encountered in their adoption.

TEACHING IN AND WITH THE COMMUNITY	The slogan "the real classroom is outside, get out into it!" does not only mean working on the school farm. It refers above all to the community and the villages in the school's catchment area. If students are to empathize with villagers (as they should), they must understand and identify with farmers' lives. Bringing outstanding community leaders to the school as resource people or hosting farmers during field days are important, but going out into their own territory is an even more important vehicle for understanding the life of farmers and their families.

Excursions

Teachers may arrange excursions to villages for many reasons:

- to let students observe or even drill a particular skill or practice;
- to expose students to a particular problem or a new idea;
- to develop group cohesion;
- to develop students' interest in other people and their communities.

Farmers' problems and outlooks cannot be understood without a study of their physical, socio-economic and cultural environment. Excursions allow teachers and their students to become an investigating team. Suitable strategies such as *Rapid Rural Appraisal* (RRA) or *Participatory Rural Appraisal* (PRA) as well as *Farming Systems Research* methods can be very useful to get to know the area around the school and, whenever there is a chance, also other regions of the country. At the same time, students are trained in observation and investigation skills and have a chance to practise these skills under realistic conditions.

Farm visits involve a complex web of interpersonal relationships. Teachers, their students and the farmers bring their own values to the meeting. Each will observe and internalize events differently. It is up to the teacher to develop an open atmosphere of interaction between students and farmers so that the latter can feel proud to be the hosts.

Farmers must be contacted before the visit. The teacher and the farmer (or other host) must agree on the purpose of the visit and its timing, while students should know what is expected from them in terms of behaviour and task performance.

The excursions should not end with the return to the classroom. What teachers and students have experienced during the excursions should serve as food for thought and discussion when the topics studied during the excursion come up on the way back to school or when the visit is reviewed in the classroom.

Student attachment

While work on the school farm can be useful, it does not come near to the real situation of working on a regular commercial farm. Restricted by its soils and climate, the school cannot grow all the important crops of the country, and a demonstration of other crops in the demonstration plot can help the theoretical courses but does not provide realistic experience. Therefore, students should get an opportunity to work for some time, perhaps between terms, on a farm, if possible even in another part of the country. While they are on a farm attachment, students should have a specific topic to look into and write a report on. These reports, in turn, can serve as important background material for an exchange of experiences among students and their evaluation by the teacher.

During their field attachment, it is important for students to live as members of the household and work alongside the farmer. Host farmers should be able to consider themselves co-trainers, and their evaluation of the students' work should be taken into serious consideration. Therefore, host farmers must know exactly what is expected from them and from the students.

Outreach programmes

In most developing countries only a small percentage of rural youth is privileged enough to attend agricultural schools and colleges. Society invests important resources in their training and so this investment should give social dividends. Students should be aware that they have a social obligation. It is not enough for them, alone, to profit from the opportunity; they must share their knowledge with the less fortunate. To do so, they need not only the right attitude towards small-scale farmers, but also suitable communication skills. One efficient way of teaching these is in outreach programmes.

While on the attachment programme, students learn from their hosts; while on the extension outreach programme, they go out to small-scale farmers in the vicinity of the school and share their knowledge with farmers who can profit from the information input.

Students might begin by participating actively in the preparation and management of field days on the school farm, and then start to work with extension specialists in rural areas. One day a week in the second or third year of training should be devoted to extension training practice in the context of the outreach programme.

To reach farmers effectively, a working knowledge of the local language is needed. This is often different from the language of instruction used in the school. This knowledge should be developed as part of the curriculum, prior to the beginning of the outreach programme. Close relations must be kept with extension workers in the area.

At the same time, the school can become a force in local development, particularly through extension activities. The outreach programme enables teaching staff to keep in close touch with the situation in the field and thus ensure that training is truly relevant and up-to-date. It can provide stimulating experiences for the teachers themselves.

Outreach programmes that have started with good intentions may often fail because some necessary conditions have not been met. Among these are a lack of resources (when no budget is provided for the programme), a rapid turnover of the teachers responsible for the programme (owing to a lack of incentives and the additional work load expected from teachers), a lack of suitable technologies to offer the farmers, clashes with local leaders who feel "left out" and the overburdening of certain communities with too many students. Teachers who conduct outreach programmes should be aware of these dangers. On the other hand, evaluations have shown positive results, especially in enhancing students' technical skills, their ability to develop good relations with farmers, proficiency in conducting demonstrations and a high interest in extension work.

BIBLIOGRAPHY

Adams, M.E. 1982. *Agricultural extension in developing countries.* Harlow, UK, Longman.

Barnes, D. & Todd, F. 1977. *Communication and learning in small groups.* London, Routledge & Kegan Paul.

Blum, A. 1976. Studies on the use of animals of economic importance in schools. *J. Biol. Educ.,* 10: 77-83.

Dodds, T. 1986. *Distance teaching for developing countries.* London, University of London Institute of Education.

Easton, G. 1982. *Learning from case studies.* Englewood Cliffs, N.J., USA, Prentice Hall.

FAO. 1990. *Make learning easier. A guide for improving educational/training materials.* Rome.

Hyman, R.T. 1974. *Ways of teaching.* Philadelphia, USA, Lippincott.

Jaques, D. 1984. *Learning in groups.* London, Croom Helm.

Minnick, D.R. 1989. *A guide to creating self-learning materials.* Los Baños, the Philippines, IRRI.

Scaife, C. & Wellington, J. 1993. *Information technology in science and technology education.* Milton Keynes, UK, Milton Keynes Open University.

Slavin, R., Sharan, S., Kagan, S., Hertz, R., Lazarowitz, R., Webb, C. & Schmuck, R. 1985. *Learning to cooperate – cooperating to learn.* New York and London, Plenum.

Teaching
aids

Whatever their style may be, good teachers use teaching aids to maximize their efforts.

The most common teaching aids are:
- textbooks and other written sources;
- non-projected visuals;
- overhead and slide projectors;
- audiovisuals;
- games, simulations and role play;
- computers;
- mnemotechniques.

TEXTBOOKS AND OTHER WRITTEN SOURCES

The most common written sources which teachers use are textbooks, teachers' guides (if available), topic summaries, extension publications, open and programmed worksheets, original or rewritten research papers and "narratives of and invitations to inquiry".

Textbooks

The textbook is still the most common written material which both teachers and students use in the teaching-learning process. If a textbook more or less covers the subject curriculum, teachers follow it rather than a curriculum guide; and for students it is the easiest aid with which to prepare a lesson or review a topic.

Unfortunately, the availability of textbooks is a major problem in the majority of developing countries. Substantial amounts of imported paper and suitable printing facilities are needed (although these are usually available for the production of newspapers). In many countries, a more serious problem is the funding, development, testing and distribution of textbooks. Although in some industrialized countries books are partially replaced by more sophisticated and more expensive learning materials, such as computer-aided instruction, in most countries books are still the most important teaching aid. Research conducted recently in developing countries has found students' success in school to be positively related to the availability of printed materials.

The lack of suitable textbooks is particularly serious in most agricultural subjects because of the relatively small readership size. Commercial publishers are reluctant to produce specialized texts and, where textbooks are produced on a non-commercial basis, agriculture usually does not enjoy a high priority. In some school subjects it is possible to use the same book over a wide range of countries where the same language is spoken but, in agriculture, textbooks must take into account geographic, ecological and even socio-economic and cultural factors. All these restrict the radius of the potential usability of an agricultural textbook.

Where textbooks are available, teachers of agriculture should be careful when choosing which chapters to use and how much to give students to read. Wherever the local situation is different from that described in the book, this should be pointed out. The textbook can be used to let students prepare for or review a topic.

In both cases, questions should be added for the purpose of checking students' understanding of the text. Teachers should be careful not to use questions suggested by the authors of the textbook without first ascertaining that students can answer them as they stand. Sometimes they have to be rephrased because the local agro-ecological situation is different from that described in the textbook.

Teachers' guides

In some cases, curriculum centres or publishers have prepared special *teachers' guides* to accompany a text. These can be of great help to teachers, although they should be used with discretion. The biggest problem with teachers' guides is that teachers often do not know they exist because bookshops are not interested in stocking them owing to the small demand and because there is often a lack of communication between the curriculum centres which produce the guides and the practising teacher in an agricultural school.

Topic summaries

If the teacher has no suitable textbook available, the next best solution is to distribute summaries of the topic. These can be reproduced in any simple way and be distributed to students at the beginning of the term or course in the form of a stapled booklet, or they can be handed out as single leaflets after each lesson and then be collected in a ringbinder.

The preparation of topic summaries is a considerable amount of work for a single teacher. Therefore, if teachers (who teach the same topic) from different schools can come together and divide the work between them, the task is easier. Each teacher can also use the topic summaries of the others. Furthermore, if this cooperative work is done in a workshop, each summary writer can get feedback from colleagues and thus the summaries may be improved.

Topic summaries are not the same as lesson summaries because each teacher will still have to plan the relevant lesson on the basis of his or her own students' prerequisites and needs. What might take up one teaching period in one case might need two in another, especially when basic conceptual requisites are missing and have to be brought into the lesson plan.

Extension publications

Publications by extension services can be the next best solution when suitable agricultural textbooks are not available. They are often good and cheap additions which update older textbooks and relate to the specific problems of a country. This is especially important where publishers distribute their books to many countries in different agro-ecological zones.

Extension publications are usually written in a language intended for extension workers or progressive farmers. Thus, their professional level and language are usually well suited for students in agricultural schools. Using these publications also makes students aware of this important extension tool, which they probably will have to use after leaving the agricultural school.

Worksheets

The term "worksheet" covers a wide range of short written forms on which students receive instructions for individualized or group work. Such forms usually leave space in which the assigned work can be finished. Worksheets can be used for different purposes, the most common of which are to:

- let students work at their own speed;
- help students who have been absent to catch up;
- give routine instructions, e.g. for laboratory work;
- help in field observations, e.g. with sketches of the observation area;
- structure the recording of data, e.g. by providing a grid or a table in which students can enter the data;
- follow up the work of students (it is easier for teachers to collect worksheets than exercise books).

Students can be grouped so that the more able help their less able peers. Alternatively, worksheets can be planned so that they become progressively more difficult. The easier worksheets will be suitable for all students while the more difficult worksheets may be given to more able students.

Worksheets are an aid to less literate students who can read the instructions but have difficulties in formulating and organizing their answers. In such a case, questions which require only a few words can be used. However, the educational goal should be to progress to questions which first require short sentence answers and then more elaborate expressions of thought.

When preparing a worksheet, special care needs to be taken to ensure a clear layout. Diagrams and other sketches can replace complicated descriptions of techniques and apparatus. On worksheets for use in the laboratory or outdoors, safety rules should be emphasized.

Sometimes worksheets contain a grid which is useful for drawing a graph, a sheet of millimetre or logarithmic paper, or any other suitable form. Figure 8 (p. 114) shows a worksheet to assess the areas on honeycombs in a standard ten-comb hive that are filled with honey, brood or pollen.

Preparing good worksheets takes time and so a team approach can make the work more efficient. If teachers from several agricultural schools who handle the same topic collaborate and divide the work among themselves, each will have to prepare only part of the worksheets to be used by all.

Programmed learning
Programmed learning was the forerunner of computer-assisted instruction, but it has the advantage that no expensive hardware is needed. The common basis for programmed and computer-assisted learning is Skinner's theory that learning is enhanced by constant feedback and

that teaching should be cut down into small units. Programmed learning is especially suitable for practising skills.

Programmed learning sheets are divided in two directions: i) vertically into a column of statements with a question or a missing element, which students have to fill in, and a column with the correct answers (see Fig. 9, p. 116); and ii) horizontally into "frames". Each frame on the right-hand side includes one statement. The smaller frame next to it

---

FIGURE 8
A worksheet with grids for indicating the distribution of brood, pollen and honey in a beehive

Name of student: .... School: .... Date: .... Hive No. ....
Colours for different areas: *brood*: .... *pollen*: .... *honey*: ....

$$S = \frac{a}{2} \times \frac{b}{2} \times \pi$$

Total brood area: .... cm²

1a	2a	3a	4a	5a
a = S =	a = S =	a = S =	a = S =	a = S =
b =	b =	b =	b =	b =

1b	2b	3b	4b	5b
a = S =	a = S =	a = S =	a = S =	a = S =
b =	b =	b =	b =	b =

6a	7a	8a	9a	10a
a = S =	a = S =	a = S =	a = S =	a = S =
b =	b =	b =	b =	b =

6b	7b	8b	9b	10b
a = S =	a = S =	a = S =	a = S =	a = S =
b =	b =	b =	b =	b =

on the left-hand side gives the correct answer to the question or the open element of the *previous* statement. Students cover the part of the sheet which they have not yet worked on. They try to answer the question posed in the statement and then move one frame down to see if the answer is correct. If it is, they continue; if not, with the help of the right answer, they try to understand why they made a mistake.

Rules for preparing programmed worksheets

1. Start with an introduction which gives the information needed and helps students concentrate on the skill to be developed.
2. Offer two or three short alternatives from which students must choose. Open questions, or "why" questions, are not suitable for programmed sheets because the space for the answer is limited and there are too many possible answers. Programmed learning is suitable for skill training and for giving students an opportunity to check themselves on specific knowledge, but not on higher conceptual categories.
3. Put only one question in a frame.
4. In the case of computation, it is advisable to show how the right answer is reached.

Research papers

Important reasons speak for the use of research reports, for example reports from agricultural experiment stations, in the teaching of agriculture. Even if students leave the agricultural school with knowledge about the latest developments in agriculture, these will soon be outdated. Hopefully, students will have received at the agricultural school tools to update themselves. One of these tools is the skill of acquiring and understanding agricultural research reports.

Through the study of research reports, students learn how new agricultural knowledge is gained and tested before useful recommendations can be released. By studying agricultural research papers, students also learn how scientific insights are turned into technologies, which have to be tested for their economic and environmental sustainability. In this way, the study of reports from agricultural research stations greatly contributes to inquiry teaching.

Research is published in two ways. First, researchers write a paper on their work as soon as possible after their investigation and publish it in an appropriate *research journal* treating the area of research to which the paper belongs. Besides important international journals, most countries also have a journal that reports on their own national agricultural research, or else their agricultural experiment stations publish the results of research as

---

FIGURE 9
A programmed learning sheet on computing the phosphor content in phosphate fertilizers

	For historical reasons, the phosphor content in fertilizers is written in the form of P_2O_5. In superphosphate (MCPM), the P_2O_5 content is 21%. Compute the pure phosphor content in superphosphate.
	The molecular weight of phosphor is 34 The molecular weight of oxygen is 16 What is the molecular weight of P_2O_5?
$2 \times 34 + 5 \times 16 = \mathbf{148}$	Based on weight, the pure phosphor content in P_2O_5 is .....%.
$(2 \times 34): 148 = \mathbf{0.46}$	Since there is 21% of P_2O_5 in superphosphate, the pure P content is only .....%.
$0.46 \times 21\% = \mathbf{9.66\%}$	If we want to know the pure P content in any fertilizer, we have to [divide or multiply?] the P_2O_5 content, which is usually indicated on the bag, by a factor of....
Multiply by **0.46** or divide roughly by **2**	

occasional papers. After enough new knowledge on a certain topic has accrued, a scientist will write a *review* to compare the different papers which were written on the topic, often criticizing their methods or conclusions. This is done in order to clear up apparent contradictions in the different papers or to come to a deeper understanding of the topic. Often, these reviews appear in special *review journals* which are published for the public at large. Thus, magazines such as *Discovery* and *Scientific American* bring reviews on all kinds of new scientific developments to readers who have a basic understanding of science.

Original research reports are not easy to read. Their language is concise and often overloaded with jargon. Therefore, students must learn to read research reports and understand why they all follow a commonly accepted format with more or less fixed sections (abstract, background, materials and methods used, results and discussion, conclusions and recommendations). To teach the skills of reading research reports, teachers can use two methods. If the report is written in awkward language, but its contents are important and interesting, then it can be *rewritten* in simpler language. If the problem is understanding the paper, a worksheet can be prepared, containing general questions such as:

- What is the purpose of the abstract?
- What was the researcher's main problem?

and specific questions such as:

- Why did the researchers use two controls in their experiment (if they did)?
- What do the researchers mean when they write on line 12, "........."?
- Why did they disagree with the previous findings of ....?

Data in research papers are summarized in different

forms: tables; curve, bar or pie graphs; photographs, etc. Students should understand the advantages and disadvantages of each of these presentations.

Narratives of inquiry

Research reports "distort" what is really happening in a scientific investigation. They provide information only on the positive, final results. The researchers may have worked for years on dozens of techniques and chemicals which did not bring about the envisaged result, yet none of this would appear in the final report. Thus, students might get the wrong idea of how research is done in reality.

A standardized research report can give students the impression that the researcher first read all there is on the topic of research, then used a ready-made method and, during the investigation stage, clearly separated data collection from analysis, leaving the interpretation for a later (discussion) stage. In scientific reports, not much can be read about chance discoveries, changes of theory and directions of investigation in the midst of the work because the first results may have shown that a completely different method of investigation would be better. Narratives of how famous discoveries have been made can correct this view and make science more human.

Invitations to inquiry

Invitations to inquiry were developed for quite a different purpose. They emphasize the planning and interpretation stages in a scientific inquiry and replace the technical work in the laboratory by providing students with the findings. Thus, many complicated research endeavours that could not be accomplished in the laboratory of an agricultural school (e.g. surgery on animals, the handling of dangerous materials, the use of expensive equipment, time-consuming laboratory operations) can be replaced by ready-made data. Real work in the laboratory often ends with unclear findings which students are unable to analyse. In an invitation to inquiry, data are so structured that they bring out the purpose of the exercise but require students to think about them.

For instance, the teacher can tell students about or present them with a written account of a case in which an insecticide has killed flies in a cowshed in the past but is now no longer effective. The teacher asks students (who have not yet learned about resistance) to hypothesize what the possible reasons for this could be. Students will probably come up with suggestions such as:

• the material was stored for too long and is no longer active;
• the spraying equipment does not work properly;
• something in the cowshed has changed;
• something in the flies has changed.

All these hypotheses could be right, in principle, and the teacher needs to be careful not to reject any suggestion without discussing it, unless it is really nonsensical. The teacher may ask how one could experimentally test each hypothesis and so eliminate the wrong ones. For instance, students might suggest trying a fresh batch of pesticide. Instead of carrying out a real experiment, which is not feasible, the teacher may tell the students that such an experiment has actually been done and that the fresh pesticide did not kill the flies either. The next suggestion is then discussed. One of the students will probably suggest using a different sprayer. "A good idea," says the teacher, and continues "This was done, but the effect remained the same." Thus, one hypothesis after another is rejected, until students suggest that perhaps something in the flies has changed. Then the next question will be: "How can that be tested?" The verbal inquiry continues until the teacher decides to tell the class about the development of resistance.

NON-PROJECTED VISUALS

Teaching aids or props attract attention. This is important because, in any classroom, there are many distractions that compete with the teacher for the students' attention. Props help the teacher to remove at least some of these distractions and to get the students' full attention. However, props do more than this. They serve as reference points for later remarks and questions. Unlike the spoken word, a prop is more tangible, and teachers and students can always come back and build on it.

Different teaching aids or props appeal to various senses. Most of them *visualize* what the teacher tries to convey verbally, often in abstract terms. Most students learn by sight (looking through their notes, reading a textbook or looking at pictures) rather than by hearing. Thus, educational teaching aids serve to concretize the abstract nature of the lecture. However, the teacher must not forget that the concrete props or examples are only a means, and their purpose is to help students conceptualize and create a picture in their mind.

Without a doubt, the best visual aid is *the real thing*. However, in most cases this is not feasible. Therefore, the next best thing is to use visuals, accompanied by sound or the verbal explanations of the teacher.

Projected visuals such as overhead transparencies, slides and motion pictures have distinct advantages – but also disadvantages, the major ones being the need for electricity and sometimes the need to darken the classroom for optimal viewing. However, there are many visual teaching aids which are independent from these restrictions, and they will be dealt with first. There are two major mistakes which teachers commonly make in the use of visuals:

• using visual aids that cannot be seen clearly by everybody; and
• overloading the visuals with too much detail.

The blackboard or chalkboard

"Talk and chalk" has often been used to describe authoritative teaching. However, this is not necessarily so. In many cases the blackboard is the only appliance a teacher has in the classroom and it is therefore of paramount importance to use it well.

The "blackboard" does not need to be black. Actually, a dark green colour makes reading easier, especially for students at the back of the classroom. There are also more expensive white boards available, on which coloured felt pens are used. These last longer than coloured pieces of chalk, do not dirty the hands of the writer and give a clearer and brighter picture.

There are "ten commandments" to optimize the potential of the blackboard (see Box, p. 122) and many of them also apply to other visual aids.

Posters and flip charts

Single, large pieces of paper can be used as posters in a similar way to a blackboard. Their advantages are:
• they can be prepared before the lesson, using large, coloured felt pens or crayons;
• they can be kept on a wall for a longer period and referred back to if necessary;
• they can be used again;
• they are easy to carry;
• sections can be covered by masked tape and revealed at the right moment;
• they can be stuck to a wall, thus progressively building up a story.

Ten commandments for blackboard use

1. Always write on a clean, dry blackboard. Clean the blackboard with a wet sponge which should be squeezed and used again. Do not write on a wet board. If you use a duster, clean it regularly by beating it lightly outside the window or classroom door.

2. Do not forget that some students have to read your writing from the back of the room. Letters should be 3 cm high, and the space between the lines no less than 1.5 cm.

3. The chalkboard is suitable for constructing a story or listing points as they come up in a discussion.

4. Important lines or points in a list can be emphasized by an asterisk, a dash or an arrow in front of it, or by using colour.

5. Use capitals only to emphasize special words or short headings. Lower-case letters are easier to read.

6. Ensure that students use the same rules when writing on the blackboard. Using the blackboard should not be a teacher's privilege.

7. Do not stand between the blackboard and the class, especially when you want students to copy from the blackboard.

8. Good teachers prepare their text (and especially drawings) before the class starts. This is especially important when you want to draw scale diagrams. It is better to prepare them ahead, cover them and expose them when required.

9. Above all, plan what you want to write on the blackboard. The area is restricted, and one of the biggest drawbacks is that you have to wipe out the previous text before you can put up the next.

10. Another problem with the blackboard is that, when writing, the teacher's back is to the class.

The size of posters for use in a regular classroom should be at least 50 x 70 cm. A poster should not contain more than seven lines with seven to ten words on each. Care must be taken to ensure clear spacing between letters and words.

Flip charts are poster sheets which are fixed together at the top so that each sheet can be flipped over the top, and thus a new chart is exposed. The advantage is that, when a sheet is turned over, it does not remain to distract students from giving their full attention to the new sheet. To use flip charts, a kind of easel support is needed.

Pin, flannel and magnet boards

Pin boards can be made of different materials which easily hold pins. The board is usually covered by a large piece of paper on to which cards of different sizes and colours can be pinned. The different cards can be reorganized on the pin board. Therefore, they are especially useful during brainstorming sessions, when students are asked individually to write down their ideas in a very concise form on coloured index cards and to put them on the pin board. During the discussion after the brainstorming session, the cards can be reorganized into meaningful structures. At the end of the lesson, the cards can be glued to the paper and thus be kept for further reference, like posters.

Pin boards are also very useful to protocol a teachers' meeting when the main ideas and conclusions are more important than taking minutes. They can be photographed and serve in this form as protocol.

Pin boards give the teacher more flexibility than a blackboard or charts. However, coloured cards can be quite expensive.

Flannel boards or flannel graphs are made of any rough-textured, neutral-coloured cloth, blanket or (preferably) flannel. This material must be pinned down tightly with clips or nails on a board, for instance a blackboard. A fixed board is preferable because it leaves the blackboard free for additional use. A flannel board is especially useful if the elements to be shown to students can be prepared beforehand. This includes pictures taken out of magazines.

Graphics should be prepared on stiff but light cardboard, with sandpaper glued lightly to the back. Pictures can be glued on the top side of the cards. Instead of being glued, the sandpaper can be stapled on to the cards. The board should be tipped back to prevent the cards from falling down. The main advantage of flannel boards is the possibility of developing a "story" by successively putting one element after the other on the board, thus keeping up action. The time during which the teacher has his back to the class is practically nil. Humorous graphics enhance the effect of the flannel board.

Magnet boards are similar to flannel boards but are more stable. Any iron or steel surface such as a metal filing cabinet will do as a board. Instead of sandpaper, small pieces of magnet are glued to the back of cards. Flannel and magnet boards can be combined by gluing a flannel surface to a metal sheet. Magnet boards can also be used outdoors.

Models and other exhibits

Models have the advantage of being three-dimensional and therefore attract more attention. This is especially true if they have moving parts. They are usually more expensive or more difficult to build than the visuals described but they are also more effective, mainly because they look more like "the real thing" they represent. For most people, they are easy to understand. Models can be smaller or bigger than the real thing, but they must be large enough for the class to see. Models can be made simpler than the object they represent, thus adding clarity. Nevertheless, they must be exact enough to convince the viewers. They should be strong and portable for handling and transport. If it is a working model, it must be easy to operate and maintain.

Other *exhibits* such as dried plants, collections of insects and other preserved animals have the advantage of being very similar to the original, but they have to be handled much more carefully than technical exhibits. Most effective conservation materials have a repulsive smell and are often toxic.

Living organisms are in many cases the real thing. From this point of view, they have the highest explanatory value. One of the huge advantages of agricultural schools and colleges with a farm attached to them is that fields and animals can be studied *in situ*, in their natural (or better, agricultural) setting. However, sometimes teachers want to keep plants and smaller animals in their classroom or in a laboratory to enhance the opportunities for students to observe them, measure developmental parameters or experiment with them (for instance, *Drosophila* flies in the study of genetics). This is especially true for non-farm animals.

Plants are usually easier to keep in the classroom or laboratory than animals, although moving animals are more attractive to most students. Keeping living animals and to some extent also living plants in the laboratory demands special arrangements such as a suitable habitat and supply of fresh food and water, including during weekends and school holidays. Many teachers find it difficult to obtain exact information on the optimal conditions for unusual animals. When schools keep experimental animals, which are usually grown in research laboratories, standard techniques, equipment and food are relatively easy to obtain, but often at a high price.

Some teachers will argue against taking wild animals out of their habitat and putting them in cages. More problematic for most are conditions which put the animals under stress. Any possibility of cruelty to the animals must be prevented.

Among the partially domesticated animals, which retain most of their natural behaviour even with human interference, bees are perhaps the most suitable to keep for demonstration purposes. This can be best done in observation hives with a glass front. Other insects can be kept for similar purposes in terrariums, and fish in aquariums. Hatching eggs in a simple incubator with a glass cover is especially attractive to younger children. Keeping small animals (usually insects) which are used in the biological control of plant pests together with these pests can provide another type of interesting object for observation and investigation.

OVERHEAD AND SLIDE PROJECTORS

Where there is a reliable electricity supply in the classroom, the *overhead projector* is the best visual aid. It has all the advantages of the blackboard and many additional ones, but scarcely any of its disadvantages. A major advantage is that teachers do not have to turn their back to students while writing or explaining overhead transparencies. They maintain eye contact with the class. Unlike other projected visuals, there is no need for a darkened room which easily puts students to sleep, especially in a warm climate. Teachers can write on a transparency as they go on teaching or they can use transparencies which they have prepared beforehand. Overhead transparencies can also be used in larger halls.

Following are some hints on how to optimize the use of the overhead projector:

- The screen base should be above the head of the audience and tilted, so that the lens and the screen are parallel and the light beam falls in a right angle on to the screen (see Fig. 10, p. 128).
- This first point is often overlooked, and the result is a distorted picture.
- Position the projector at a distance from the screen that enables the screen to be filled.
- Prevent bright light shining on the screen (which can be any white surface, although a commercially manufactured screen gives clearer pictures). This should also be done when the room is lit artificially.
- Make sure the picture is sharp. Adjust the focus.
- Use a slim pointer (e.g. a knitting needle or pencil – nothing thicker) to indicate the part of the image that relates to your explanation. You can place the pointer on top of the transparency to prevent any nervous shaking of the hand. Do not turn your back to the class in order to point to the image on the screen.
- Do not conceal the image. Stand to the side of the projector.
- When there is text on the transparency, give students time to read. Never talk about something different while students read the text. Give them 20 to 30 seconds to read. During that time you can observe your students and think how to continue.
- Do not use more than 10 to 12 transparencies in one lesson.
- Never move an overhead projector while the lamp is still warm. Most projectors have a fan to cool down the delicate lamp.
- Have a reserve lamp ready in case the one you use burns out.

• The rules for preparing script on overhead transparencies are similar to those for posters:
  - the script should be at least 6 mm but better 8 mm
    high;
  - there should be a maximum of seven lines with no
    more than seven to ten words on one line;
  - use strong colours such as black, red, blue or dark
    green – other colours should only be used when they
    have a specific meaning;
  - always check your transparencies on the projector
    before using them in your class.

In order to make overhead transparencies more effective,
a sheet of paper can be used to expose the message
progressively. Thus, the eyes of the students are focused
on the specific point being explained. This can even be a
transparent paper through which the teacher can see what
is to come. In this way, overhead transparencies can serve
the teacher as lecture notes.

Parts which are no longer needed can be covered by a sheet
of paper, thus focusing the viewers' attention on the exposed
text or picture.

Transparencies can be made more sophisticated by adding

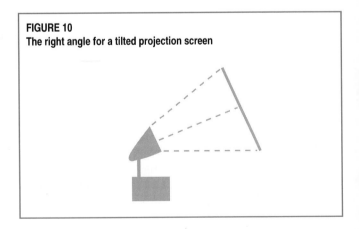

FIGURE 10
The right angle for a tilted projection screen

overlays in contrasting colours. This is especially useful when explaining how organisms, machines or other systems are composed of elements which then add up to a more complex whole.

Most photocopying machines enable copying from books and other written sources. When using this option, take care not to overload the transparency. Most tables in publications should be simplified. Concentrate on the most important data.

Slides

Slides have the advantage of showing small objects enlarged and in their natural colours. They are also easy to carry and store. Their biggest disadvantage is the need for blackout conditions which only few classrooms have. The best slides are those taken with a camera. Unless teachers are photographers, it is difficult for them to prepare their own slides. They usually have to buy them or copy those of a colleague. Such copies are expensive and often of a low quality. The slide projector is probably at the back of the room and the teacher must therefore employ one of the following techniques:

- use a well-functioning remote control, which most schools do not have;
- stand behind the class and lose eye contact in the darkened room; or
- request help from a projectionist.

In any case, care should be taken not to prolong a slide show.

AUDIOVISUALS

A slide show can be combined with recorded sound, for instance with an *audiotape*. In such a presentation, the sound (text, music or other sounds such as those of the animals shown) and the duration of the picture are coordinated. The advantage is that the show proceeds smoothly, appealing both visually and aurally. However, in most cases teachers prefer to give their own commentary and to allow students to ask questions. This is difficult in the midst of a slide-tape show.

Audiotapes alone are less effective than when they are combined with slides. However, they are easy to prepare, especially when they are recorded from a radio transmission. They are most helpful for dyslectic students.

Motion pictures and video cassettes

Motion pictures have the advantage of being able to show *action*. They can condense or stretch time – an important factor when following the development of plants and animals. Motion pictures are rapidly being replaced by video cassettes which have several advantages over motion pictures and scarcely any of their disadvantages. Video cassettes are much easier to operate and do not need a completely darkened room as motion pictures do. Video equipment is becoming cheaper and even the filming of video cassettes is much simpler and also becoming cheaper. Moreover, video films can be borrowed from libraries. Many schools have a video recorder which allows one-time occurrences such as pest damage in a field to be photographed. The event can then be shown to students who may not be able to get out in time to observe the phenomenon first hand in the field.

The main problem with any kind of instructional film is that students remain passive during the show. Therefore, a film sequence should not be longer than 15 to 20 minutes. If it is longer, the teacher should make a break in the middle and allow students to ask questions or to discuss issues shown in the film.

EXERCISE GAMES, ROLE PLAY AND SIMULATION GAMES

There are different and mostly overlapping definitions for role play and different kinds of games which are used for teaching purposes. However, they have a basic feature in common: they create situations in which students learn something while being engaged in a playful activity, without necessarily being aware that they are being taught.

There are two typical points regarding games: i) they have an objective, which is usually for a participant to win, and hence there is usually a competition between individuals or teams; and ii) they have rules, which are simpler than in "real life". In order to win, students have to apply known strategies or develop new ones. They have to make decisions in simulated situations.

Many psychological arguments have been used to justify and encourage the use of games for learning in schools. The most basic one is probably that this is the natural way in which children have always learned skills and social rules in all cultures. Children observe adults and imitate them in their play. Adults also learn to do new things through role play, and they enjoy this exploratory learning. With simulation games it is possible for students to learn and have fun at the same time.

For most learning situations, the teacher must first motivate the students. This is not so in a learning game. Since students are already motivated to play, and since learning is required to play and succeed in a learning game, students are ready to learn. This is especially true for students who are less motivated or less able. The latter often have a chance to do as well as their more gifted peers owing to the luck factor which is nearly always present in a game. At the same time, simulation games provide bright students with the opportunity to develop alternative strategies which help them to win. Such students enjoy developing their own strategies within the rules of the game.

Students also learn that much in life does not depend on their efforts. Unpredictable events affect our lives constantly. This is another reason why many games have a built-in luck factor in the form of chance cards which have to be drawn or dice which have to be thrown, for instance.

Simulation games stretch the attention span of students because the interaction between them serves to keep the purpose of winning before them and thus prolongs their motivation.

Simulation games and role play provide students with an opportunity to participate actively. In role play, they are compelled to adapt to new situations. Students do not conceive them as preparations for later life, like much of school learning, because in order to win or collect points, the play situation has to be mastered *now*. Thus, the distance between learning and applying is eliminated, unlike much learning in school.

Different types of thinking are encouraged by simulation games; above all analytical and intuitive thinking. Games reward analytical thinking because the analysis of one's own and the other players' moves and the consequences of these moves are necessary for succeeding in the game. Simulation games are a good exercise for critical "if A, then B" type thinking.

On the other hand, students are also encouraged to try intuitive thinking, which is quicker but also more risky than analytical thinking. They can do this because the penalty for failing is limited to the game and thus the risk is not considered real. In the debriefing of a simulation game, students can evaluate which of the two ways of thinking – the analytical or the intuitive – is better suited. They learn that decision making is not such an easy task.

In simulation games in which decisions have to be made, students can make bad decisions. They find that out either by losing in the game or by coming to the conclusion during the debriefing that they should have employed a different strategy. They learn this by themselves, without the teacher having to tell them what mistakes they made. This adds to the relaxed atmosphere which is typical of simulation games. Yet, in many simulation games and role play the teacher can influence the game by unobtrusively changing the situation and thus forcing students to adapt to the new circumstances.

While many games encourage competition, they can also be constructed in such a way that cooperation is the only way to win. In role play and simulation games, students take up specific roles with which they learn to identify.

The moves of one player affect those of the others. Thus, an interdependency is created between students. They learn to observe each other's behaviour and also to improve their communication skills.

Different types of games induce different types of learning

1. *Exercise games* foster recall of details or train students in the application of rules. Their rules are relatively simple and there is usually *only one right solution.*
2. *Simulation games* have more complicated rules. They involve decision making for which different strategies can be used. Students usually need more time to play them.
3. *Role play* (as defined here) emphasizes the interaction between players and trains them to adapt their strategy of action when the situation changes. *The solution is usually open.*

There is no clear line which differentiates simulation games and role play. In both simulation games and role play, affective objectives such as the fostering of interests, attitudes and values are usually quite important, as are students' behavioural patterns, arguments and decisions. Therefore, the teacher should explain the character of the simulation before it begins. Even more important is the debriefing session after the game, which will end with a discussion on what students think they have learned from playing the game.

The following sections present examples of games with an agricultural or scientific background.

Exercise games

Exercise games are usually designed on the basis of card or board games which students know well. The simplest are quartets in which players have to collect four cards belonging to the same series. They soon learn which specific card fits into a generic group. Such games can be easily adapted to give students an opportunity to recall chemical elements that have the same valency, four varieties of a crop species, weeds that belong to the same botanical family and so on. As in rummy, the one who manages to get rid of all his or her cards first wins.

Other card games let students exercise their knowledge of scientific rules. For instance, in the *Formulon* game, students draw cards with chemical symbols and an indication of whether they are anions, cations or elements with all their electrons. In this case, the series of cards to be deposited in order to win are chemical compounds which actually exist and which have to be composed from ion and atom cards, observing the rules of valency. Competing players will point out mistakes made by other students who try to put down chemically impossible "compounds", using unsuitable cards. There is even a "joker" in the game, called Mendelieff.

In the *Predator* game, each card shows a plant or animal, with a list of what it eats and what eats it. Different games can be played to demonstrate rules in the web of life. The Death and Decay cards make the game more sophisticated.

The *Fungus Life Cycle Game* uses a board and differently formed cards, which show the symptoms of crown rust (*Puccinia coronata*) during different seasons, microscope pictures of the different spores which belong to the sexual and asexual life cycles and markers to show when the fungus changes its host. The cards must be placed on the empty spaces printed on the board. The game was developed after an analysis of college students' difficulties in understanding how the sexual and asexual cycles interchange and which symptoms in the field match the different stages of spores studied under the microscope. An evaluation exercise showed that such a game actually can improve students' grasp of a complex biological development which has to be understood in order to protect cereal crops from a major pest.

Simulation games

Simulation games are more complex because they try to simulate a situation in nature or in social life – although of course in a very schematic way, as the following examples will show. They are also more difficult to construct than exercise games. Experience and various trial runs are needed before the learning and messages built into the game actually "work" in the way the designer has planned.

Extinction is a game which trains students in the understanding of ecological processes. It is played on a game board which depicts the imaginary island of Darwinia. Students can place their "population cubes" on different habitat types, and concepts such as ecological density, population dispersion, population growth, competition, predator-prey interaction, mutualism and many more come up. With the help of different cards (e.g. gene cards, rates of reproduction, natural and human-caused barriers), conditions are changed and the plants, represented by the players, must adapt. (There is a danger that less able or younger students might accept the anthropomorphic view that "plants decide what they should do to survive").

The aim of the *End to Hunger* game is to indicate the way in which ecological, social and technological factors are liable to affect the ability of a developing nation to provide its population with the necessary supplies of food, while showing how mutual cooperation and aid in development are likely to improve the lot of the individual nations and, thus, of the world as a whole. Each player represents an unnamed nation. A few players start, by chance, with a better economic situation. In the beginning, all players have to rely on manual labour (i.e. writing a short sentence as quickly as they can). They are also affected by chance cards. Depending on their financial situation, they can obtain education and buy a "food machine", and thus progress; but this is difficult without cooperation. After each round (representing a season), students have to decide what to do with their gain or loss. Players record their actions on a Development Balance Sheet, which allows them to analyse in the debriefing phase of the game how rational or irrational their moves have been.

Monsoon is a simulation game which focuses on some of the dynamics of poverty in an imaginary village near Mysore, India. Players take the roles of farmers and experience some of the forces that shape people's lives in a village. Other major roles are those of the moneylender and the development worker. Among the issues raised are dependence on the monsoon, cooperation and leadership within the village, the polarization of rich and poor, the role of the moneylender, the relevance of development programmes, the role of the village development worker, the effect of irrigation, malnutrition and its consequences and the pressures of social conformity.

Other simulation games, for example *WEEDING*, are based on computer models. Biological processes and constraints to management are simulated, based on real weather data. This makes it possible to practise decision making about weed control under semi-realistic conditions. The game can also serve as an introduction to the complexity as well as the risk and uncertainty of pest management decisions. The main disadvantage is, of course, the need to have access to a computer.

Role play

In most role plays, a certain basic situation is assumed and students are asked to play the roles of representatives of groups which have different interests but need to reach a common decision. The outcome depends to a large degree on how well the players are prepared and how persuasive they can be in presenting their arguments and propositions. Often in role plays, especially on development issues, some of the students form a committee while others represent the public at large. This type of role playing is good training for democratic behaviour and being ready to listen to opinions one is opposed to.

In all role plays, students are presented with some facts of the situation and the problem they are expected to face. The roles are usually already defined (although, in some cases, this too is left for students to decide). However, the amount of realistic background facts with which students must work can differ considerably. If few facts are available to students, they have to be more inventive. But in such a case the discussion may also become less realistic.

Three main levels of preparation are possible in role playing. These can be described using the case of the *Super Tom* role play. In this play students are told that a new tomato variety, Super Tom, has been developed in the country. It is especially suited for the manufacture of ketchup, and the Minister of Agriculture has high hopes of turning it into an export success. However, Super Tom is susceptible to a serious plant disease and plant protection is costly. The Minister appoints an Agricultural Development Committee, which consists of six members: the Permanent Secretary of the ministry as chairperson, the competent pest control officer, a rural economist and a representative of the farmers, the consumers and the food manufacturing industry. The committee is asked to advise the Minister if Super Tom should be recommended for growing on a large scale.

At the most basic level of sophistication, the teacher only nominates the simulated committee and explains the situation and the task facing the committee. In such a case, the risk is that nearly all members quickly agree with the recommendation and no useful discussion can develop. This can be prevented at the second level of sophistication, by giving each member of the committee a brief which he or she is to follow, at least in the beginning of the deliberations. Thus, for instance, the chairperson and the representative of the industry are strongly in favour, while the representative of the consumers, who fears for the supply of fresh tomatoes, and the pest control officer, who knows about the big pest control problem (and has, like others, personal grievances), are definitely against Super Tom. Between these positions is the farmers' representative, who wants to know if the prices are fixed and who will supply the fungicides. The economist always points out what the different arguments mean in economic terms. Thus, much in the discussion depends on the strength of the arguments brought forward.

The teacher can change the situation quite inobtrusively by playing the role of a clerk who brings the committee fax messages which change the situation. For instance, a fax message could report that the stores in Europe and the United States have a surplus of ketchup which will last for the next five years. This will immediately strengthen the committee members who are against Super Tom, and all the players will have to review their position. The next message could change the situation in the opposite direction, for example an offer by a large country to buy the whole production volume at a high price.

At the third level of sophistication, every member of the committee has access to a wealth of background material, as in a case-study (see the section Case-studies, p. 89). In the case of Super Tom, this would include the research reports on the development of Super Tom, data on the field trials, production calculations, the vitamin content of fresh tomatoes and ketchup and so on. At the case-study level, the learning effect grows, and students who make good use of the background material have an advantage. On the other hand, the part played by oratory declines. Preparing role plays at this level requires a lot of work.

COMPUTER-ASSISTED INSTRUCTION

Microcomputers can be very useful tools in education when they are used efficiently. However, their purchase and maintenance are expensive. Therefore, their cost-benefit ratio must be weighed against that of other instructional means. In this section, it will be assumed that students have at least limited access to microcomputers for instructional purposes.

Classroom teachers cannot be expected to develop their own computer programs (or *courseware*, as instructional software is often called). Therefore the availability of suitable courseware is the most crucial issue. When there is a choice, the first decision in the selection of courseware often concerns the degree of *designer control* versus *student control*. In the latter case, students have more possibilities to choose the level of difficulty, the rate of progress and the use of computer tools suited to them from the courseware's menu. These programs are also referred to as being *menu-driven*. Teachers, and even more so students, usually have less knowledge of the pedagogical advantages of different sequences than programmers assume.

Most of the early educational computer programs were drill-oriented. They followed the patterns of *programmed learning* (see the section Textbooks and other written sources, p. 109).

Today, courseware is developed with a much wider pedagogical scope. Following are six categories of instructional computer programs which differ mainly in their educational goals. As is the case with most classifications, the borders between the different categories are not always clearly fixed.

1. Drill and practice. These programs give learners immediate and specific feedback. Questions and task assignments are usually structured in such a way that the learner ticks off one of several possible answers, and the program can react with a "right" or "wrong" statement. More sophisticated courseware recognizes typical mistakes and switches the learner to remedial sections in the program. These can be skipped by advanced learners, thus speeding up their learning process.

2. Tools. These programs train learners how to master software which is also used in "real life". The most typical of these tools are spreadsheets, databases, word processors and spelling checks, graphic utilities, expert systems and statistical packages. With the exception of word processors, these tools usually employ little text. Compared with other tools, therefore, they are less dependent on language ability, can be easily translated into other languages and are less culture-specific. This is not necessarily true for the courseware that introduces tools to the learner. These software tools are largely user-controlled, can be used in many subjects and for purposes which were not foreseen by their developers. Microcomputers are being used more and more in farm management.

3. Simulations and modelling. Unlike computer tools that make older systems more effective (e.g. word processors instead of typewriters and filing systems), simulations are used when the real experiences they simulate are too dangerous, costly or time-consuming, or impossible in a given school or training situation. Thus, simulations are often presented instead of laboratory exercises. They are also helpful when the real events are too fast or too slow to be observed by students. They are useful for training students in handling variables and displaying data. Simulations have been found to be useful as training tools and are also used as *simulation games* (see the section Teaching in the laboratory, p. 94).

4. Computer literacy and programming. Computer literacy usually means a generalized understanding of the place of computers in our lives, their potentials and limitations and our skill in their use. As in non-computerized literacy and numeracy, there are different levels of attainment:

i) *Novices*. Students use ready-made courseware in which exact instructions are given about which keys have to be used to proceed to the next frame. This stage is typical for most computer-assisted instruction.
ii) *Intermediate level*. Literacy includes the use of software tools such as word processors, spreadsheets, information retrieval systems or application of analytical computer packages.
iii) *Advanced level*. Computer users have some programming skills.
iv) *Professionals*. Users have attained a level permitting their employment in the data processing field.

5. *Communication and networking.* Some schools in industrialized countries use the computer to communicate via electronic mail with operators at other stations and with databases. Advocates of this type of computer use emphasize its motivational value. Using electronic mail in itself does not necessitate many skills. In most cases, the advantages of electronic speed do not balance the costs incurred by modems and extra telephone lines.

6. *Computer-steered instrumentation.* This use of computers is becoming increasingly important in the training of workers in automated industries in which machines are computer-steered. Simple computers are also used in agriculture, for example in the monitoring of irrigation – where water use is critical – and in the monitoring of optimal conditions in greenhouses.

The most important microcomputer tools for use in agriculture are spreadsheets, databases and expert systems, which can be of great help in farm management. Training is required for their efficient use, however. Although microcomputer courseware can be very useful in teaching how to keep farm records and how to make good farm management decisions, the role of computers in agricultural training programmes occurred at a late stage. The main problems encountered with computer-assisted instruction in agricultural education are unsuitable hardware systems, lack of pretested courseware and negative teachers' attitudes, which are often coupled with insufficient teacher training in the use of microcomputers.

In the United States, the majority of vocational-agricultural teachers indicated in a survey that they use computers in their work. However, it appears that in many cases the main use was in telecommunications. In Europe, computer education has become an important part of training for agricultural technicians and extension workers. So far, too few evaluation studies have been made on computer-assisted instruction in agriculture to enable valid judgements on its effect on students.

The development of courseware for agricultural education is still in its infancy. Courseware is usually developed by commercial enterprises or curriculum centres. Agricultural schools and colleges are too small a consumer group to become a worthwhile target for commercial coursework designers, and curriculum centres with an ability to produce good agricultural courseware are rare.

Computer courseware must be adapted to changing agro-ecological and socio-economic conditions, like other curriculum materials. In agriculture, site specificity is even more important than in the basic sciences. The best solution is probably to create regional teams of agricultural teachers and software programmers who will adapt suitable courseware to their own situation, as a first step until they have gained enough experience to embark on original courseware creation. Suitable handbooks and regional clearinghouses, in which information stored on CD-ROM can be downloaded to diskettes or paper output, should be of considerable help to such courseware development groups.

The overall aim should be to familiarize students with the potentials of the microcomputer for farm management and other uses in agriculture (e.g. expert systems) and to introduce them to the software which is used in the agricultural sectors of different countries.

Heat, dust, humidity and vibrations are more of a hazard to microcomputers in tropical and semi-arid countries than elsewhere. In developing countries, power often fluctuates more than most computers can tolerate, and technical devices to overcome this problem are often unavailable or expensive, as are batteries. Other specific problems in developing countries are a lack of suitably trained hardware and software operators for servicing and repairing equipment. In many countries, a certain computer hardware system and its compatibles might be more common than other systems for which more courseware has developed.

Especially in developing countries, clear priorities should be set for computer-assisted instruction. The first priority should be the training of agricultural researchers, trainers and extension leaders in universities. Agricultural colleges, and where possible agricultural schools, should train their students in the use of adapted software, especially spreadsheet-based farm management programs and relevant courseware. This will enable students, after leaving school, to use programs within the framework of an extension service, both for the generation of advisory packages and for administrative purposes. Once a certain degree of computer literacy exists, microcomputers can also be used for desktop publishing of locally produced instructional materials on specific agricultural topics. This could at least partially relieve the chronic lack of suitable textbooks.

Computer-assisted instruction and the use of computer tools is certainly not the first need to be met in agricultural education and extension. However, the prices for hardware are falling constantly and computer software is becoming more and more user-friendly. In time, the computer will also find its place in agricultural education. However, one has to be careful not to invest in handsome "gadgets" before other more pressing needs are met.

MNEMOTECHNICAL PHRASES	Mnemotechniques are used to remember pieces of information, especially details, when they cannot be embedded into a logical framework. Although the main purpose of an *acronym* is to shorten a long name, it can also serve to remember the original name. We often invent a story around the digits of a telephone number in order to remember it. Also in agriculture, such mnemotechniques can easily be created by teachers and students. Following are some examples of how students remembered properties of grape varieties using mnemotechnical phrases:

"Little green pearls make a good New Year present" (the Perlette variety has small green berries and ripens at the beginning of the season).

"The cardinal in his red robe walks in front of the procession" (the Cardinal variety has big red berries and ripens early in the season).

"Chocolate mousse is eaten at the end of the meal" (the Black Muscat variety ripens towards the end of the season).

BIBLIOGRAPHY   **Alessi, S.M.** 1985. *Computer-based instruction – methods and development.* Englewood Cliffs, N.J., USA, Prentice Hall.

Barlex, D. & Carre, C. 1985. *Visual communication in science.* Cambridge, UK, Cambridge University Press.

Blum, A. 1993. Microcomputers for agricultural education and training in developing countries. *In* FAO. *The potentials of microcomputers in support of agricultural extension, education and training.* Rome.

Davison, A. & Gordon, P. 1978. *Games and simulations in action.* London, Woburn.

Farrell, J.P. & Heyneman, S.P. 1989. *Textbooks in the developing world. Economic and educational choices.* Washington, DC, World Bank.

Gerlach, V.S. & Ely, D.P. 1980. *Teaching and media – a systematic approach.* Englewood Cliffs, N.J., USA, Prentice Hall.

Glazier, R. 1976. *How to design educational games,* 6th ed. Cambridge, Mass., USA, ABT Associates.

Horn, R.E. & Cleaves, A., eds. 1980. *The guide to simulations/ games for education and training.* London and Beverly Hills, Sage.

Miles, R.S., ed. 1982. *The design of educational exhibits.* London, George Allen and Unwin.

Mugglestone, P. 1986. *Planning and using the blackboard.* London, Heinemann Educational.

Rogers, A. 1986. *Teaching aids.* Milton Keynes, UK, Milton Keynes Open University.

Taylor, J. & Walford, A. 1972. *Learning and the simulation game.* Milton Keynes, UK, Milton Keynes Open University.

Van Ments, M. 1983. *The effective use of role-play. A handbook for teachers and trainers.* London, Kegan Paul.

Wilkinson, J. 1979. *The overhead projector.* London, British Council Media Department.

Zachmann, R. & Siri, C. 1993. *Visual presentation of research results.* Ibadan, IITA.

References to the simulation games mentioned in the text

Blum, A. 1976. A game to teach the life cycle of fungi (The Fungus Life Cycle Game). *J. Biol. Educ.,* 10: 203-207.

Blum, A. 1978. *Super Tom.* Rehovot, Israel, H.U. Faculty of Agriculture.

Blum, A. 1979. The remedial effect of a biological learning game. *J. Res. Sci. Teach.,* 16: 333-338.

Blum, A., Elazar, E., Feinstein, B., Mittleman, S. & Shulman, Y. 1975. *End to Hunger.* Jerusalem, Ministry of Education, Curriculum Center.

Chemical Teaching Aids. *Formulon.* Letham, Ladybank, Fife, UK.

Hubble, S.P. 1971. *Extinction.* Stamford, Conn., USA, Sinauer Associates.

Lowell, M. 1973. *Predator – The Food Chain Game.* Oakland, USA, Ampersand.

Staley, J. 1981. *Monsoon.* Bangalore, India, SEARCH.

Wiles, L.J., Wilkerson, G.G. & Coble, H.D. 1991. WEEDING: A Weed Ecology and Economic Decision-making Instructional Game. *Weed Technol.,* 5: 887-893.

Monitoring and evaluating students' learning

The term *monitoring* usually describes a situation in which a process is supervised or controlled while it is going on and in such a way that immediate remedial action can be taken. The expression is more often used in technology. For instance, monitors are common in intensive care units of hospitals, where they sound an alarm to inform doctors when one of the critical physiological processes in a patient falls below a certain minimum.

In education too, monitoring means "holding one's finger on the pulse" of students' learning. Thus, monitoring is *a kind of ongoing evaluation* in which the remedial action the teacher or instructor can take is more important than a grade. It emphasizes the educational value of evaluation over the all too common notion that evaluation means examining and delivering a final judgement.

Evaluation should cover not only knowledge (and other cognitive achievements), but also skills and attitudes. The methods of attainment of *cognitive* objectives, including intellectual skills, are usually tested in the classroom. *Technical skills* must be assessed where they can actually be performed – in the laboratory or in the field. How far students' *attitudes* and *interests* have changed and how far they have internalized the values taught cannot be "tested" in an examination. Students would most probably give the expected answers and not necessarily what they really feel and believe. However, behavioural changes, which can be observed by good educators, can give an indication of students' attitudinal learning.

Self-evaluation is more common in training programmes for adults, who would object to formal examinations of the type set in school. However, self-evaluation can also serve a useful purpose in agricultural schools, above all to raise students' awareness of the extent to which their subjective self-assessment is similar to or different from the objective assessment of the teacher or external examiner.

By using different monitoring techniques, good teachers know quite accurately what their students do or do not know long before examination time.

Monitoring techniques

Unless the class is too big, good teachers attempt to involve *all* students in a verbal *dialogue*, and not only those who are quick to put up their hands. Questioning students without creating an examination atmosphere is the best way to monitor students' learning.

Another monitoring measure is to collect periodically *students' workbooks* and other products of their independent work. When *worksheets* are used, students can continue working on new worksheets when one is with the teacher. This is more difficult with workbooks.

From time to time, a third monitoring device can be used: short *quizzes* of two to three questions which all students answer in writing. These are mainly questions on knowledge and understanding and should not take more than ten minutes to answer.

Monitoring is important for teachers because it enables them to take immediate remedial action. It is also a help to students because it makes them aware of their current performance.

Verbal versus written examinations

Verbal and written examinations both have their advantages and disadvantages. *Verbal* examinations offer an advantage for students who may have difficulty expressing themselves in writing. From the teacher's point of view, they allow him or her to probe more deeply into the understanding of students, who in a written examination may produce only general statements. Verbal examinations pose two main problems: how to ask all students equally difficult questions; and how to minimize the emotional tension created by the "one-one" situation. In both cases, it is difficult to be objective, which is a cause of fear for students.

There are some rules, however, that can be followed when preparing verbal examinations (see Box, p. 154).

Written examinations are not free from biases either. They are often called "paper and pencil tests" and examine only things that can be answered with these tools. Written examinations nearly always have a high correlation with the reading and writing ability of students, which is not pertinent to the subject being tested. Often we do not only test knowledge of a certain subject, but also to a large degree how "test wise" a student is. This is more the case in multiple choice questions than in essay examinations, but in the latter case there can be much less objectivity in grading. No examination is perfect.

Constructing test items and the taxonomy of educational objectives

Before deciding on the form of a test, the objectives of the examination must be clear and correspond to the learning objectives. An examination that tests the knowledge of facts while the curriculum unit attempted to teach intellectual skills (e.g. the application of principles to the solution of problems) will not do. It would test students' ability to remember facts learned by rote, and not their ability to use this knowledge, as intended by the objective.

Improving verbal examinations

1. Prepare the questions in writing and grade them according to their difficulty or (even better) according to their taxonomic category (as explained later in this section). This shows already that a good verbal examination takes as much preparation as a written one.

2. Put sets of comparable questions into different envelopes and let students draw one of them. Another way is to disclose all the questions ahead of time and thus give students a clear cue as to what is expected of them. They can then draw lots on who will answer the different sets of questions. This distribution of questions among students gives them a feeling of fairness.

3. Start with some easy questions to "break the ice". This will take away much of the anxiety which nearly always exists in a verbal examination.

4. Beware of external circumstances which can distort the impression you receive in a verbal examination. A mediocre student who is examined after a very bright student will make a worse impression than if he or she comes after a very weak student. The external appearance of a student can influence the examiner in a verbal examination more than in a written one (although other irrelevant factors such as the handwriting of the student can affect the examiner in the latter case). The only preventive measure against these dangers is to be aware of them.

The most common system for putting objectives and test questions into categories according to what they test is Bloom's Taxonomy of Educational Objectives. The categories in this taxonomy are *knowledge, understanding, application, analysis, synthesis* and *evaluation*. These categories are arranged in a hierarchical order. To answer correctly a question which belongs to a higher level, one must be able to master a relevant question in the lower categories. For instance, students might *know* a formula because they have learned it by heart but this does not mean that they can *understand* the meaning of it, much less apply the formula when needed. Vice versa, in order to apply a rule correctly (e.g. to set up a control treatment in an experiment), one has to *understand* the principle behind this rule and, to do so, one has to *know* it.

Unfortunately, most examination questions ask for knowledge, while only a few test understanding. Application items are rare, although they are the most crucial in agriculture, and all agricultural institutions emphasize how important it is for their students to be able to apply knowledge in their future work. For each examination, teachers must decide how many questions of each category they want to include in the test, according to the objectives of the curriculum and the way they taught it.

Types of test items

After selection of the content areas and after a review of the objectives, the most suitable types of test items can be chosen. The first decision is usually whether the test should consist of *essay questions* or *objective test items* (or a combination of both). Of the two, essay questions can probe deeper into students' intellectual skills such as critical and analytical thinking, organization of thoughts, power of expression, ability to expose an idea concisely and even creativity. Essay questions favour bright and able students. Objective items are better for testing factual knowledge. Teachers often prefer a mix of the two in order to test a wider scope of abilities, but also to give intermediate-level students a fairer deal. Such mixed examinations take more time and are therefore best suited for end-of-term examinations.

Essay questions. This type of test item can ask students to: describe a phenomenon, for example the symptoms of a plant disease (knowledge); explain a biological process, for example nitrification (understanding); apply a scientific rule in a new situation (application); analyse a complex farm situation on the basis of documents (analysis); plan a solution to a farm problem, for which knowledge from different areas is needed as input (synthesis); or evaluate the implementation of a land consolidation programme. The higher the taxonomic level of the question, the more time and intellectual skills are needed. Essays take longer to grade, and there is a danger that subjective factors may influence the grading. Therefore, in crucial tests such as diploma examinations, at least two examiners should grade each paper on the basis of agreed criteria.

Questions such as "what do you know about X?" are unclear and encourage students to write a lot without going into depth because they cannot be sure what the teacher expects. Knowledge questions are better asked in the form of objective items.

Objective items. Although these items can be quickly marked and graded, it takes time to construct a good item. Therefore, item banks in which teachers of a given subject pool their test items are very helpful. Two examiners marking an objective test independently should arrive at the same grade. In most objective items, students have only to indicate the right answer and are not required to express an idea. They test students' knowledge rather than their ability to write.

The most common objective test items are *multiple*

choice questions in which the student has to indicate the right answer out of four or five suggested statements. The others are called *distracters*. Writing four distracters (besides the right answer) instead of three lowers a student's chances of guessing the right answer from 25 to 20 percent. However, it is not usually worth while to work on a fourth distracter because it is difficult enough to construct three good ones which are not obviously wrong.

Multiple choice items can go beyond pure knowledge. They can contain different explanations or reasons for a process. To answer such a question, students have to understand what is going on in that process. In many questions in science and agriculture, it is easier to construct a multiple choice item in which several answers are right and to ask students to indicate the *wrong* answer. In such a case, the words *wrong* or *not true* must be emphasized.

When writing multiple choice items, one must take care not to give hints as to which is the right answer. Correct answers should be no longer than the distracters. Generally, the main information should be contained in the question, the "stem", and the alternatives from which to choose should be short.

The best distracters are those that represent typical mistakes made by students. Such items have a high diagnostic value.

True/false or *alternative-response items* are in principle multiple choice items with only one distracter. Students have only to indicate which of the two statements is the correct one. There is a 50 percent chance of guessing the right answer. In most cases, the fact that true/false items are more easily constructed does not justify the high chance of guessing. Of course, one can also add two other basically possible answers: "both statements are right" and "both are wrong".

In opinion polls, true/false is replaced by *agree/disagree* items. However, students should not be graded on their opinions! This would lead to insincere answers.

Matching is another form of multiple choice questioning, consisting of two columns of words, short sentences, pictures or signs. The purpose is to test students' ability to identify the relationship between two things. Students are asked to draw a line from one item in the first column to the corresponding item in the second column, or to write next to each item in the first row the number which indicates the matching item in the second column. Writing more items in the second than in the first column or indicating that a right-hand item can be matched to more than one left-hand item makes answering by elimination and guessing more difficult. The columns should not contain more than seven to ten items and should be written on the same page.

Completion items require students to insert the right word, phrase or number into a given sentence. Either two or three options are suggested from which the student must choose – for example, "the amount of concentrates fed to a milking cow depends on .... (her body weight/her daily milk production/both together) – or no alternatives are proposed. In the first case, the completion sentence is like a multiple choice item, but more awkward to answer. In the second case, additional answers are possible, for instance "how much the farmer can afford". The teacher may not have expected this answer, but is it wrong? Therefore, completion sentences should only be constructed with a number of alternative answers indicated, or where only one exact answer can be correct; for example, "the boiling point of pure water at sea level is ..... °C".

Grading. Only in a few cases do the grades given to a student in an examination correspond to the number of right answers. Let us assume that students passed a very easy test with high marks or failed in a very difficult exam. Does this necessarily mean that they have achieved mastery in the first case or that they have learned nothing in the second case? It could well be that the difficulty level was not adequate and that the marks should therefore be adjusted according to a standard. The two most common standards are *norms* and *criteria*.

In a *norm-referenced* test, teachers compare students' achievements with a statistical norm. They grade students "on the curve"; that is, they assume that the distribution of achievement in the class is "normal". Therefore, there will be as many students with low grades as students with high grades in both of the tests described above. This may take the sting out of the tests that were too easy or too difficult, but it does not show to what extent the students have achieved the objectives of the course. This is exactly what criterion-referenced tests aim to achieve.

MONITORING AND EVALUATION IN THE LABORATORY

Most of what has been said about monitoring and evaluation in the classroom is also true for the same activities in the laboratory and in the field. There are, however, some differences. In the laboratory, teachers can *observe* students while they actually perform an activity. If a student does not use the microscope in a correct way, the teacher sees it and can correct the mistake on the spot.

The best way *to monitor* students' understanding in the laboratory, besides talking with them while they work, is by *laboratory reports* in which students not only report the results they have obtained but also interpret their meaning. Sometimes report forms also contain further questions to probe students' understanding.

Evaluation of students' learning in the laboratory depends on the purpose of the laboratory work. If this is to acquire laboratory *skills,* then these must be evaluated by testing how far students are able to demonstrate these skills, for example how accurately and perhaps also how quickly they can perform a learned activity. Technical skills should not be evaluated by paper and pencil tests.

However, written tests can be useful to evaluate students' understanding of *how to plan laboratory experiments.* This is especially important in guided discovery or inquiry teaching (see the section Teaching styles, p. 74). In such a test, students are provided with background information on a problem that has to be solved. They are then asked to prepare a written plan of how they propose to set up a laboratory experiment to solve the test question. Before students actually implement this plan, they hand it over to the teacher or examiner, who checks if the proposed procedure can lead to the answer. The grade given to this part of the examination will depend on the ideas expressed and not on the feasibility of setting the experiment up in the laboratory. It may be that a student's ideas are good but that the local laboratory equipment or the time available does not allow the ideas to be carried out.

If the students' ability to perform standard tests in the laboratory is also to be tested, the teacher will now exchange each student's proposals with a standardized instruction sheet that takes into account the equipment that is ready to be set up or is already in place (depending on the objective of the test and the time available). Thus, the two parts of the examination – the experimental planning and the laboratory skills – are evaluated separately.

MONITORING AND EVALUATION IN THE FIELD	On the one hand, monitoring students' learning *in the field* is made more difficult by the complex nature of many agricultural activities but, on the other hand, the teacher (or instructor) can devote more time to the *observation* of

an individual student, because the student-teacher ratio is usually smaller in the field and students work for a longer period on a given job.

It is important to check that each student has performed all the activities he or she is supposed to learn often enough so as to be able to perform them at a standard level, as demanded by the learning objectives. For this purpose, it is advisable to prepare a record booklet in which all the skills to be learned are listed.

Such a list usually has additional columns in which the instructor can indicate the dates of when a student has first performed a job, when he or she has reached a standard (or minimal) level of performance, and other important facts for monitoring (e.g. difficulties to be overcome) (see Fig. 11).

Such a continuous record is better than a one-time field examination which can never realistically test out-of-season jobs (e.g. picking fruit after the picking season). Therefore, *formal examinations of practical objectives* should be distributed over different seasons. If the evaluation is performed by an outside examiner (e.g. a

---

FIGURE 11
Example of a page in an objective-based job record booklet

Objective	First performed	Reached standard level	Observations
1. Prepare mechanical sprayer			
2. Prepare spraying mixture according to instructions on the label			
3.			

regional extension officer), students get to know him or her, and thus examination anxiety is minimized. Such an arrangement also has the advantage of strengthening links between the school and extension services. As a fringe benefit, extension officers will often be able to exchange their experiences with those of the school staff.

In these field examinations, the proficiency level of students can be observed, and they can be questioned on their understanding of why a given job should be performed in one way and not in another. The evaluation can be based on the observed process, or on the final product (if such a product can be made during the evaluation period). Besides qualitative *proficiency, efficiency* can also be tested, if it was among the set learning objectives. There must be agreement among examiners, teachers and instructors as to what level of performance should be expected, and students must be informed accordingly.

An evaluation of students' performances during *outreach* activities and while on a *farm attachment* should be part of the final grade.

BIBLIOGRAPHY

Black, H.D. & Dockrell, W.B. 1986. *Criterion-referenced assessment in the classroom.* Edinburgh, UK, Scottish Council for Research in Education.

Bloom, B.S., Hastings, J.T. & Madaus, G.F. 1971. *Handbook on formative and summative evaluation of student learning.* New York, McGraw-Hill.

Ebel, R.L. 1979. *Essentials of educational measurement.* Englewood Cliffs, N.J., USA, Prentice Hall.

FAO. 1991. *Improving training quality. A trainer's guide to evaluation.* Rome.

Lewis, R. 1984. *How to help learners assess their progress: writing objectives, self-assessment questions and activities.* London, Council for Educational Technology.

Nash, R. 1973. *Classrooms observed. The teacher's perception and the pupil's performance.* London, Routledge & Kegan Paul.

Popham, W.J. 1975. *Educational evaluation.* Englewood Cliffs, N.J., USA, Prentice Hall.

Rowntree, D. 1977. *Assessing students – how shall we know them?* London, Harper & Row.

Satterly, D. 1989. *Assessment in schools.* Oxford, UK, Blackwell.

Teacher and instructor training

This book is meant primarily for practising teachers. Therefore, we shall not discuss how teacher training institutes should be. However, the sad fact is that many teachers of agriculture begin their teaching career without proper pedagogical training. They are often seconded by a Ministry of Agriculture to an agricultural school. Even if they have attended one or several courses on agricultural education, this is certainly not enough. The situation is even more serious with instructors on the school farm, who have an important educational role to play but are considered by many authorities as having only a technical task.

Therefore, this chapter is divided into two parts: *Initial training* of teachers and instructors and their *continuous training*. Since both teachers and instructors should basically be educators, we shall treat their training together and differentiate between them only where their functions differ considerably owing to the main scene of their activities: the classroom, the laboratory or the field.

INITIAL TRAINING OF TEACHERS AND INSTRUCTORS

Agricultural teachers and instructors in agricultural colleges and schools are selected and employed mainly on the basis of their professional agricultural training and experience. This might not be deep enough, but should be better than the training and experience of other candidates. Therefore, weak points should be strengthened through continuous in-service training. As mentioned above, it often happens that beginner agricultural teachers or instructors have not received any formal pedagogical preparation. Therefore, they need initial training when beginning their job.

Everyone has had some *non-formal* or *informal experience of teaching*. This might have been as a peer teacher, when helping a friend to come to grips with a piece of new information or with an unfamiliar skill. It could also have been as coach in a sport team, as a leader in a youth movement, as an extensionist in an agricultural service, or even as a parent. This experience, coupled with an inborn educational instinct, can give beginner teachers a valuable set of positive, pedagogical "entering behaviours". Often, reflecting on these experiences can help the beginner to build on them.

Beginner teachers tend to teach the way they were taught. If they were lucky and their teacher was a good educator, by imitating his or her teaching style, they can develop a positive approach to teaching, even if their repertoire of teaching skills is restricted. However, more often than not, beginner teachers imitate the average lecturers at their agricultural training institutions who *lectured* and did not consider themselves to be educators.

The beginner teacher or instructor who has not received preservice training should at least attend an induction training course and/or be coached by a tutor during the first year of teaching.

Induction training

The school, or the school system to which it belongs, should organize a concentrated *induction course* before the beginning of the school year. In this, mainly practical teaching skills should be exercised, accompanied by explanations of why and when to use them. At this stage, there is not usually much time for theory. However, beginner teachers should become familiar with good, practical reference books, which they might want to consult later, while accruing experience.

The main thrust in an induction course should be on suitable teaching styles, while a move should be made away from lecturing. The emphasis should be on teaching with practical examples, on the use of a variety of teaching aids (and especially the less common ones), on inquiry training and problem solving and on close cooperation with the school farm and the community. The latter is especially important to assure that the values of the community are not offended.

Most new teachers come from outside the region and are not familiar with the local community. Therefore, they should get to know it from the beginning. Probably the best way is to let new teachers conduct a *Rapid Rural Appraisal* in one of the villages in the catchment area of the school, choosing a central issue which is connected with the new teacher's field of specialization.

Tutorial training

Another form of induction training consists in the *tutoring* or coaching of each novice teacher by one of the best teachers in the school. From time to time the tutor might visit a lesson given by the new teacher, and discuss how to improve the novice's teaching. Tutors must not behave like inspectors. Their role is that of an experienced and empathetic friend. The tutor should visit the new teacher's class only if the latter feels comfortable in that situation. The tutor will invite the new teacher to visit his or her own class and to discuss the methods used during that lesson. At the same time, they will analyse the behaviour of various students together in order to enhance the new teacher's understanding of their problems and how to deal with them.

Similar procedures can be employed in the induction of a new *field instructor*. In this case, methods used in the classroom are less important. It is more crucial to achieve an educational approach. The school farm should be run on commercial lines to teach students how a farm is managed and also because most schools need the income from the farm, but this must not prevent instructors and field managers from acting as educators. In the final analysis, the school farm is there for the students, and not vice versa. An experienced field instructor with an educational approach should tutor and coach the field staff who come into contact with students.

CONTINUOUS TRAINING OF TEACHERS AND INSTRUCTORS

Lifelong learning is important for everyone, especially for teachers and instructors. The forms can be varied, e.g. self-directed reading, participating in a distance learning scheme, attending training sessions, working part-time for a higher degree, working on a project, cooperating with researchers and so on. Some of these forms might not be well developed in a particular country, but opportunities to improve exist everywhere for those who show initiative.

In the case of agricultural teachers, continuous learning should cover three areas: i) advances in agricultural science and technology; ii) educational innovations and a more in-depth understanding of students; and iii) leadership qualities.

Updating and upgrading agricultural knowledge

Agricultural science is probably the quickest changing of the above three areas. It demands a constant watch for new developments. This does not mean that teachers should include everything new in the curriculum, but they should be familiar with the developments in their respective fields of specialization. The school must keep in its library at least the *agricultural journals and magazines* which are published in the country (in many countries there are not many). The same is true for the reports of regional experiment stations. Unfortunately, these are often missing in agricultural school libraries.

Another good way for teachers to keep up with new developments is to look into the *publications* which are *published by and for the extension services*. Agricultural teachers can gain by attending training sessions and visiting experimental demonstration plots which are arranged for agricultural officers in extension and other agricultural services. Such meetings are often held not far from the school and do not necessarily involve a great expense.

Educational in-service training

Although there are not so many new developments in education which a school can afford, regular in-service training on educational topics is a must. If teachers are not exposed to new ideas, they easily fall into a routine which chokes new initiatives. Thus, teaching easily becomes boring for students and teacher alike, and teachers soon get tired of their job.

The same is often the case with field instructors. Some of them come to see students as a nuisance which distracts them from what they see as the main task: to run the farm smoothly and without being disturbed by students. Yet, while teachers' in-service training in educational matters is widely accepted and often encouraged by incentives, the same is not very common for farm instructors – quite wrongly so.

Leadership training

Teachers of agriculture and trained instructors can and should fulfil leadership roles in their school and in their community. This is usually an additional burden to their work. Therefore, they should be encouraged to take on public tasks and receive credit for it. They should also be given an opportunity to develop their leadership skills in suitable training institutions.

Incentives

Some teachers feel the need to progress and to "recharge their batteries" by looking for further training opportunities, even if these demand sacrifices. Others need incentives. In both cases, schools should be interested in encouraging teachers to use in-service training opportunities. These incentives could be in different forms: part-time release from other duties, subsidies for the costs involved, advancement in their job, public recognition and the like. In some countries, teachers get seniority additions to their salary only if they participate in in-service activities during school vacations. Unfortunately, many civil service regulations, especially in developing countries, do not provide financial incentives for teachers to continue their training and do not encourage teachers to work for a higher qualification. The quality of agricultural education depends on the quality of its teachers and instructors, whose performance improves with continuous training. Therefore, it is in the highest interest of any school to encourage and help its teachers and instructors to participate in in-service training.

BIBLIOGRAPHY

Cropley, A.J. & Dave, R.H. 1978. *Lifelong education and the training of teachers*. Oxford, UK, Pergamon.

Dove, L.A. 1986. *Teachers and teacher education in developing countries*. London, Croom Helm.

Dow, G. 1979. *Learning to teach: teaching to learn*. London, Routledge & Kegan Paul.

Ethirveerasingam, N. 1990. Teacher education in agriculture. In *Training for Agriculture and Rural Development 1989-1990*. Rome, FAO.

FAO. 1991. *Improving training quality. A trainer's guide to evaluation*. Rome.

FAO. 1992. *Planning for effective training. A guide to curriculum development*. Rome.

Mills, H.R. 1986. *Teaching and training: a handbook for instructors*, 3rd ed. London, Macmillan.

Thomas, E., Sharma, M., Khanna, A. & Jatoi, H. 1993. *Policy and practice in initial teacher training.* London, Commonwealth Secretariat.

UNESCO. 1973. *Technical and vocational teacher education and training.* Paris.

Curricular aids for teachers of agriculture

Teachers of agriculture who want to develop their own curriculum, especially in developing countries, have no easy task because their access to good examples is limited. So far, there are no depository libraries for curricular materials in agriculture.

An additional problem is the lack of textbooks and other curricular materials which are suitable for tropical and semi-arid conditions. However, sometimes examples of agricultural curricula or other materials that help to devise curricula can be found. The most common are described, with examples, in the following sections.

Monographs on curriculum development in agriculture

Such publications are rare. Examples are:

- Raman, K.V. & Sudarsanam, R. 1991. *Curriculum development for higher education in agriculture.* Rajendranagar, Hyderabad, India, National Academy of Agricultural Research Management.
- Olaitan, S.A. 1984. *Agricultural education in the tropics: methodology for teaching agriculture.* Basingstoke, UK, Macmillan.
- Stonehouse, D.P. 1995. Curriculum development for the achievement of multiple goals in the agri-food industry. *Eur. J. Agr. Educ. Ext.*, 2(4): 1-14.

Curriculum recommendations in reports on conferences and workshops

When national or regional conferences or workshops on agricultural education are held, reports are usually published which contain useful ideas and recommendations, but no "blueprints" for curriculum development. Examples are:

- *Proceedings of the First Workshop on Curriculum Development.* Lilongwe, Malawi, Forest Training Programme, 1991.

- *Proceedings of a Workshop on Agricultural Curricula for Undergraduate Students.* University of Dar-es-Salaam and Ford Foundation, 1976.
- *Regional Seminar on Agricultural Curriculum Development.* Nouméa, New Caledonia, South Pacific Commission, 1972.

Descriptions of innovative educational institutions and activities
Ideas for curriculum planning can also be found in reports on innovative institutions that have developed their own curricular approach. Examples are:

- Bawden, R.J. 1992. Systems approaches to agricultural development: the Hawkesbury experience. *Agric. Syst.,* 40: 153-176.
- Boukli, N. & IBE staff. 1975. *Mostaganem Institute of Agricultural Technology: an educational innovation.* Paris, UNESCO.
- van den Bor, W., Wallace, I., Nagy, G. & Garforth, C. 1995. Curriculum development in a European context: an account of a collaborative project. *Eur. J. Agr. Educ. Ext.,* 2(1): 1-16.

The latter is a critical evaluation of a collaborative project under the EU's TEMPUS programme involving the Faculty of Agriculture, Debrecen Agricultural University, Hungary, the Department of Agricultural Education, Wageningen Agricultural University, the Netherlands, and the Agricultural Extension and Rural Development Department, University of Reading, United Kingdom.

Descriptions of actual curriculum planning and implementation
These are reports on actual curriculum development efforts, from which other curriculum developers can glean ideas for developing their own curriculum. Examples are:

- FAO Programme for Better Family Living. 1974. *Curriculum Development Workshop for Bukura Institute of Agriculture*. Nairobi, FAO/PBFL.
- Bochnert, J. 1988. *Agroforestry and agricultural education, with a focus on the practical implementation*. Gaimersheim, Germany, Verlag Josef Markgraf.

Guidelines for curriculum content

Sometimes, but not often, very detailed guidelines on how to choose the content for a specific subject area are available. An example is:

- FAO/USAID/UP. 1981. *Food, nutrition and agriculture. Guidelines for curriculum content for agricultural training in Southeast Asia*. Los Baños, the Philippines.

Syllabi

Syllabi with short descriptions of the objectives and more detailed lists of topics and practicals are very useful. Often, the number of periods which should be devoted to a topic are suggested. Such syllabi are usually published by national authorities or institutions. Examples are:

- Ogongo Agricultural College. 1992. *Diploma and certificate programmes*. Windhoek, Namibia, Ministry of Agriculture, Water and Rural Development, EC Agricultural Training Project.
- Botswana Agricultural College. 1976. *Syllabus for certificate in agriculture*. Gaborone, Government Printer.

Curricula with more details

Less often, agricultural education institutions publish more detailed curricula, with indications of what teaching and learning materials might be used. One of the most extensive, which was adapted by several other agricultural institutes, is the following:

• FAO Programme for Better Family Living. 1974. *Bukura Institute of Agriculture. Vol.1 – Curriculum for farm management, home economics, extension and engineering;* and *Vols 2 and 3 – Curriculum for crops department and animal production department.* Nairobi, FAO/PBFL.

Some curriculum guides specify competency-based, behavioural objectives; topics and contents; the allotment of periods for different topics; and detailed indications of teaching and learning strategies and aids. An example is:

• *Curriculum for the certificate in agriculture at the Highlands Agricultural College.* 1978. Papua New Guinea, Agricultural Education and Training, Department of Primary Industry.

Manuals and guides for teachers

In other cases, the authorities who have published an official syllabus later release more specific materials for teachers. Examples are:

• Botswana Agricultural College. 1977. *Course manual: certificate in agriculture.* Gaborone, Government Printer.

• Botswana Agricultural College. 1988-1991. *Course details for the certificate in agriculture* (drafts). Gaborone, Botswana Agricultural College.

Sometimes, checklists for practical skills are combined with a column in which the teacher can write assessment remarks. An example is:

• *Continuous assessment book – vegetable production* (also available for other production areas). Gaborone, Ministry of Education, Agricultural Education.

In the case of primary school agricultural education, when pupils are not at an advanced level of English, teachers' manuals are even more important. Such books leave it to the teachers to prepare their own lessons, but give them details on content and methodological suggestions. An example is:

- Bergmann, H. & Butler, R. 1980. *Primary school agriculture. Teacher's manual* (with five parts: pedagogy, crops, farming methods, crop storage and land tenure). Eschborn, Germany, GTZ.

Teachers' notes on matriculation examinations

Such publications are especially welcome when a school prepares students for matriculation examinations in agriculture. An example is:

- Teachers' notes for O-level. 1983. *Zimbabwe secondary school agriculture study guide modules.* Harare, Ministry of Education and Culture.

Learning materials for students

In the best case, there are textbooks available that have been developed especially for students in a particular country or region and published by a government body or a commercial publisher. Examples are:

- Owen, G.H. 1974. *Agricultural science.* Lusaka, Longman Zambia.
- Schools Agricultural Panel, Ministry of Education, Swaziland. 1977. *Modern agriculture.* Cape Town, South Africa, Oxford University Press.

The *Agricultural Science Materials Project* of the Australian Curriculum Development Centre published a series of booklets in which two cartoon "farmers", Betty and Bruce, ask students many questions. In order to give an answer, students are obliged to think.

Students' books which were written for a science curriculum sometimes contain useful information, especially in biology, which is very relevant to agriculture. Examples are:

- Project Group No. 12-16. 1985. *Milk*. Enschede, the Netherlands, Stichting voor de Leerplanontwikkeling.
- Alabi, R.O., Asun, P., May J.B., Ndu, F.O.C. & Ndu, L.O. 1982. *Integrated science for junior secondary schools*. Lagos, Longman Nigeria.

Extension materials

Even when textbooks are available, good extension pamphlets can add actuality and introduce students to the kinds of publication they will have to work with in future as extensionists or in similar positions. An example is:

- FAO Better Farming Series, e.g. *Upland rice* (No. 20). The first 26 volumes in this series were based on the agricultural training course prepared by the Institut africain de développement économique et social (INADES) for use by extension workers. The later volumes have been prepared by FAO for use in agricultural development at the farm and family level.

Extension materials often contain more pictures than words so that even poor readers can follow the instructions. Examples are:

- Agricultural Extension Training Centre, Monrovia, Liberia. 1974. *How to grow rice*. Rome, FAO.
- Government of Botswana. *Dryland crop production – using an animal-drawn tool carrier.* 1974. Gaborone, Agricultural Information Service.

Resource books and centres

Resource books and lists of educational resources published by agricultural education institutions are important sources when working on a new or revised curriculum. Some of these publications list resources and contact addresses from which teachers of agriculture can obtain publications or other materials. An example is:

- Rural Studies Group. 1982. *Agriculture in your school curriculum. Book 3: Resources.* Carlton, Australia, Education Department of Victoria, Curriculum Services Unit.

Especially in developing countries, teachers find it more difficult to obtain help from commercial and public companies. They will often need to develop *do it yourself* instruments. Here again, a suitable manual can help. An example is:

- Community Development Counselling Service (compiler). 1980. *Remote areas development manual.* Washington, DC, Peace Corps Information Collection and Exchange Center.

Glossary

Achievement test
A test that measures the extent to which a person has acquired knowledge or skills as a result of learning.

Advance organizer
Use of already learned mental structures to teach new information.

Affective domain
The affective domain includes interests, attitudes, values and feelings.

Assessment
The process of measuring or estimating the value of something, often students' learning. Assessment is often used as a synonym for evaluation.

Audiovisuals
Materials which reinforce the reception and recall of messages which are perceived through hearing (audio) and/or seeing (vision) (see the sections Non-projected visuals and Audiovisuals, p. 120 and p. 129).

Authoritative (teaching)
A teaching style in which the teacher dominates, gives orders, allows students to ask questions only at fixed points in the lecture and expects students to remain passive (see the section Teaching styles, p. 74).

Auto-instruction
Learning with the help of books and other media, without the assistance of a teacher.

Behavioural objective

A statement that describes what skills students are expected to have acquired through learning. The skills, which can be observed or measured, may be in the cognitive, affective or psychomotor (practical) domain.

"Black box"

An instrument is a "black box" when it is used by a person who can operate it but does not know how it functions inside. The same term is applied when people use a concept without knowing all about it. In our lives, we use many "black boxes".

Brainstorming

Participants in a group are encouraged to express creative ideas, even "wild ones", on a certain topic. All these ideas are immediately recorded and only later discussed.

Case-study

An instructional method in which students are provided with information on a real-life situation and are taught how to analyse it so that they can generalize from it or draw other conclusions from the case.

Cognitive domain

The domain of knowledge and what people can do with it.

Compensatory education

Education given to students to overcome physical or mental disabilities or difficulties.

Concrete-operational stage

According to the child psychologist Piaget, this is the developmental stage in which children operate with concrete concepts and are unable to abstract them. They are unable to deal with possibilities which are not before them or which they have not yet experienced.

Coordinated curricula
Curricula in different subjects which are prepared by separate curriculum developers who interact during the curriculum development process.

Cost-benefit analysis
The ratio (numerical relation) between the costs to produce something and the benefit derived from it.

Criterion (plural: criteria)
A standard to judge a performance, e.g. in a test. A test is criterion-referenced if the achievement measured or assessed is compared with a criterion, e.g. the educational objective.

Curriculum (plural: curricula)
Guidelines for an educational programme. It usually consists of a syllabus (containing the objectives to be achieved, the contents to be learned, methods of teaching, the time allotted to different topics) and the means (learning and teaching materials) by which the objectives are expected to be reached. In the broadest sense, it is all the planned experiences that students are exposed to in order to achieve the learning goals.

Decision tree
A method to describe graphically a series of decision points at which one of two or more alternatives has to be chosen.

Decoding
Translating a coded message into an intelligible one which "makes sense".

Didactics
The theories of instruction (pedagogy) which deal with methods that improve teaching in real situations.

Disciplines, hierarchy of
The theory that some scientific disciplines are built on more basic ones, e.g. chemistry on physics, and physics on mathematics.

Discovery learning, guided
A method by which students are guided to discover facts and generalizations by themselves through inquiry, and with the help of different sources, but without the teacher or the book giving them the "right answer" straight away.

Distance education
Students learn by themselves, guided from a distance by correspondence, radio, television and other media or a combination of these. Often students are in charge of their own goals and speed of progress.

Effectiveness
A measure of the extent to which an activity reaches its objective.

Efficiency
"Effectiveness divided by time". In education: the degree to which the outcomes of teaching relate to the inputs in terms of time, costs and human effort.

Encoding
Converting a body of information into a form that can be communicated.

Essay item
A question in a test which has to be answered by writing a complete thought in one or more sentences.

Evaluation
The use of data from various sources and by various methods to judge the value of a person, a product or a process, usually in relation to a criterion or norm. The term is often used as a synonym of assessment.

Examination
A series of tests, usually taken at the end of a learning/ teaching cycle or as a condition to obtain certification of one's achievements.

Experiential learning
Students learn by "experiencing" a soft or non-fixed system in which they participate actively. The wholeness of the system is considered more important than the sum of its parts (see the section Experiential learning, p. 33).

Farming systems approach
An approach to the study of farm problems in which the farm and wider units such as villages are seen as interdependent systems. The problems of the farm cannot be understood or solved by looking at single elements alone. A sequential, farmer participatory approach to generate, evaluate and disseminate agricultural technology.

Flow chart
A graphic description of a series of processes and their interactions.

Formal-operational stage
The stage in Piaget's developmental psychology in which students are able to operate on hypotheses and deduce potential relationships, which can be verified by observation or experiments. At this stage, they can express abstract thoughts.

Gender issues
Issues in which the different roles of men and women are recognized, e.g. their respective roles in managing a household.

Induction training
Practical training which is given to a person taking up a new job. It is directed almost completely towards the specific demands of the new position (in-service or on-the-job training).

Instructional objective
A statement that describes what is expected from learners at the end of the instruction.

Integrated curricula
Curricula in which subject matter from different disciplines is used in order to emphasize higher levels of organization, e.g. science as a more universal concept than biology, chemistry and physics.

Interactive teaching
A teaching style which invites students' active participation, interaction and even initiative.

Invitation to inquiry
A guided discovery method (see the section Textbooks and other written sources, p. 109).

Knowledge
Acquired information (facts and generalizations) which can be recalled. Pieces of knowledge can be learned by rote, without really understanding them. Therefore, the Taxonomy of Educational Objectives differentiates between knowledge and comprehension.

Learning activities
Learning activities are planned to facilitate students' learning.

Learning experiences
Learning experiences are what students get out of learning activities.

Mnemotechniques
Methods to improve the power to retain information in the memory.

Monitoring
Observation of a process, and its regulation while the process goes on.

Motivation
The drive or need to do something, e.g. to learn.

Narrative of inquiry
A method of guided discovery learning (see the section Textbooks and other written sources, p. 109).

Norm

A measure or standard for the evaluation of students' test results. In norm-referenced tests, students' achievements are compared with those of a larger group.

Objective test

A test in which the person who checks the answers has no influence on the score. Typical objective tests contain multiple choice and similar items in which there is only one correct answer.

Participatory Rural Appraisal (PRA)

A Rapid Rural Appraisal (see separate entry) in which the participation of farmers/peasants is emphasized.

Prevocational agriculture

Agricultural education at an early stage, before students choose agriculture as a career and train for it.

Programmed learning

A teaching/learning method in which the assignments, the information needed to answer the questions and the way to do this are provided to learners in a structured form of writing or as a computer program. Students progress at their own speed.

Psychomotor domain

The domain which deals with physical skills in which body movements are directed by the brain. In agricultural education, this term is used to describe practical skills.

Rapid Rural Appraisal (RRA)

A method to find out in a relatively short time meaningful information on the situation in a rural region. It is cheaper, quicker and often more useful than a formal survey, and better than a problematic "quick visit".

Reductionism
The method in science that advocates the separation of factors in an experiment in order to prevent one from interfering with the others. It is also the simplification of complex data and information.

Role play
Students play roles in a simulated situation, with the aim of getting a better feeling for the problems arising out of the situation which is simulated. Role play is used to train in purposeful interaction.

Simulation games
Games in which complex life situations are reduced to a more simple form so that students can get a feeling for the simulated situation. Simulation games are similar to role play, but they have usually more defined rules and use devices (cards, boards, etc.) which are typical for games.

Skill
A manipulative or intellectual operation performed with learned competency.

Socratic style
A dialogue in which the teacher directs a student towards a truth. The teacher serves as an example for the student. The philosophy and educational method of Socrates (c. 470-399 BC) was described by his student Plato.

Spiral curriculum
A curriculum in which students are expected to return to a

given topic at a later stage and at a higher level, after they have been exposed in the meantime to additional prerequisites for the higher level.

Standard of performance
i) The measure of how well a job, a skill or an assignment is performed; or ii) that part of a behavioural objective which describes how well a task must be performed.

Standardization
Creating equal conditions for all students during a test.

Syllabus (plural: syllabi)
The part in a curriculum which describes the objectives and contents of an educational programme and often also the methods to be used as well as the expected outcome.

Taxonomy (of educational objectives)
A taxonomy is a classification which assumes a certain hierarchy among the classes it describes. The classes in the cognitive domain of Bloom's Taxonomy of Educational Objectives are: knowledge, comprehension, application, analysis, synthesis and evaluation.

Teaching skills
The "things" a teacher is able to do to enhance or facilitate learning. Examples of teaching skills are: enhancing motivation, effective verbal and non-verbal communication with individuals and groups, using different types of questions.

Teaching style
Teaching styles express a teacher's basic approach to teaching. The most common teaching styles are

authoritative, interactive, Socratic and guided discovery approaches.

Team teaching

A form of instructional organization in which two or more teachers are cooperatively responsible for the planning, teaching and evaluation of a group of students.

Test

Part of an examination, usually designed to measure cognitive, affective or manipulative abilities and achievements.

Transfer of learning (or training)

The use of knowledge, insights and skills acquired in one situation or subject to enhance learning in another situation in which the earlier learned facts and generalizations can be applied to solve new problems.

Tutorial training

Individualized training with the help of a teacher or instructor.

Vocational agriculture

Agricultural training to prepare students to enter agriculture as a career or to improve knowledge and skills in agriculture.

Index

Index

A

Achievement test 187
Adult education 57
Advance organizer 11, 19, 21, 22, 24, 44, 64, 65, 81, 187
Affective domain 42, 187, 188
Affective (skills) 15, 134, 197
Affiliation 15, 46
Agrarian countries 47, 48
Agricultural knowledge systems 5, 6
Agricultural production 5, 58
Animal production 56, 57, 180
Application (learning) 23, 24, 26, 27, 29, 82, 133, 142,
    153, 154, 155, 156, 196
Application of science 7, 51
Applied social science 57
Assessment 45, 69, 152, 162, 163, 180, 187, 191
Attachments 104, 105, 162
Audiovisuals 13, 62, 109, 129, 187
Authoritative (teaching) 73, 74, 75, 79, 121, 187, 197
Auto-instruction 187

B

Behavioural objective 180, 188, 196
Biological (topics) 20, 28, 30, 49, 51, 59, 60, 90, 95, 126,
    135, 137, 147, 156
Blackboard 13, 67, 121, 122, 123, 124, 126, 146
Black box 19, 21, 95, 188
Brainstorming 123, 188
Buzz session 88

WHERE TO PURCHASE FAO PUBLICATIONS LOCALLY
POINTS DE VENTE DES PUBLICATIONS DE LA FAO
PUNTOS DE VENTA DE PUBLICACIONES DE LA FAO

• **ANGOLA**
Empresa Nacional do Disco e de
Publicações, ENDIPU-U.E.E.
Rua Cirilo da Conceição Silva, N° 7
C.P. N° 1314-C
Luanda

• **ARGENTINA**
Librería Agropecuaria
Pasteur 743
1028 Buenos Aires
Oficina del Libro Internacional
Alberti 40
1082 Buenos Aires

• **AUSTRALIA**
Hunter Publications
P.O. Box 404
Abbotsford, Vic. 3067

• **AUSTRIA**
Gerold Buch & Co.
Weihburggasse 26
1010 Vienna

• **BANGLADESH**
Association of Development
Agencies in Bangladesh
House No. 1/3, Block F, Lalmatia
Dhaka 1207

• **BELGIQUE**
M.J. De Lannoy
202, avenue du Roi
1060 Bruxelles
CCP 000-0808993-13

• **BOLIVIA**
Los Amigos del Libro
Perú 3712, Casilla 450
Cochabamba;
Mercado 1315, La Paz

• **BOTSWANA**
Botsalo Books (Pty) Ltd
P.O. Box 1532
Gaborone

• **BRAZIL**
Fundação Getúlio Vargas
Praia do Botafogo 190, C.P. 9052
Rio de Janeiro
Núcleo Editora da
Universidade Federal Fluminense
Rua Miguel de Frias 9
Icaraí-Niterói
24 220-000 Rio de Janeiro
Editora da Universidade Federal
do Rio Grande do Sul
Av. João Pessoa 415
Bairro Cidade Baixa
90 040-000 Porto Alegre/RS
Book Master Livraria
Rua do Catete 311 lj. 118/119
20031-001 Catete
Rio de Janeiro

• **CANADA**
Le Diffuseur Gilles Vermette Inc.
C.P. 85, 151, av. de Mortagne
Boucherville, Québec J4B 5E6
UNIPUB
4611/F Assembly Drive
Lanham MD 20706-4391 (USA)
Toll-free 800 233-0504 (Canada)

• **CHILE**
Librería - Oficina Regional FAO
Calle Bandera 150, 8° Piso
Casilla 10095, Santiago-Centro
Tel. 699 1005
Fax 696 1121/696 1124
Universitaria Textolibros Ltda.
Avda. L. Bernardo O'Higgins 1050
Santiago

• **COLOMBIA**
Banco Ganadero
Revista Carta Ganadera
Carrera 9ª N° 72-21, Piso 5
Bogotá D.E.
Tel. 217 0100

• **CONGO**
Office national des librairies populaires
B.P. 577
Brazzaville

• **COSTA RICA**
Librería Lehmann S.A.
Av. Central
Apartado 10011
San José

• **CÔTE D'IVOIRE**
CEDA
04 B.P. 541
Abidjan 04

• **CUBA**
Ediciones Cubanas, Empresa de
Comercio Exterior de Publicaciones
Obispo 461, Apartado 605
La Habana

• **CZECH REPUBLIC**
Artia Pegas Press Ltd
Import of Periodicals
Palác Metro, P.O. Box 825
Národní 25, 111 21 Praha 1

• **DENMARK**
Munksgaard, Book and
Subscription Service
P.O. Box 2148
DK 1016 Copenhagen K.
Tel. 4533128570
Fax 4533129387

• **DOMINICAN REPUBLIC**
CUESTA - Centro del libro
Av. 27 de Febrero, esq. A. Lincoln
Centro Comercial Nacional
Apartado 1241
Santo Domingo

• **ECUADOR**
Libri Mundi, Librería Internacional
Juan León Mera 851
Apartado Postal 3029
Quito

• **EGYPT**
The Middle East Observer
41 Sherif Street
Cairo

• **ESPAÑA**
Mundi Prensa Libros S.A.
Castelló 37
28001 Madrid
Tel. 431 3399
Fax 575 3998
Librería Agrícola
Fernando VI 2
28004 Madrid
Librería Internacional AEDOS
Consejo de Ciento 391
08009 Barcelona
Tel. 301 8615
Fax 317 0141
Llibreria de la Generalitat
de Catalunya
Rambla dels Estudis 118
(Palau Moja)
08002 Barcelona
Tel. (93) 302 6462
Fax (93) 302 1299

• **FINLAND**
Akateeminen Kirjakauppa
P.O. Box 218
SF-00381 Helsinki

• **FRANCE**
Lavoisier
14, rue de Provigny
94236 Cachan Cedex
Editions A. Pedone
13, rue Soufflot
75005 Paris
Librairie du Commerce International
24, boulevard de l'Hôpital
75005 Paris

• **GERMANY**
Alexander Horn Internationale
Buchhandlung
Kirchgasse 22, Postfach 3340
D-65185 Wiesbaden
Uno Verlag
Poppelsdorfer Allee 55
D-53115 Bonn 1
S. Toeche-Mittler GmbH
Versandbuchhandlung
Hindenburgstrasse 33
D-64295 Darmstadt

• **GHANA**
SEDCO Publishing Ltd
Sedco House, Tabon Street
Off Ring Road Central, North Ridge
P.O. Box 2051
Accra

• **GUYANA**
Guyana National Trading
Corporation Ltd
45-47 Water Street, P.O. Box 308
Georgetown

• **HAÏTI**
Librairie «A la Caravelle»
26, rue Bonne Foi, B.P. 111
Port-au-Prince

• **HONDURAS**
Escuela Agrícola Panamericana,
Librería RTAC
El Zamorano, Apartado 93
Tegucigalpa
Oficina de la Escuela Agrícola
Panamericana en Tegucigalpa
Blvd. Morazán, Apts. Glapson
Apartado 93
Tegucigalpa

• **HUNGARY**
Librotrade Kft.
P.O. Box 126
H-1656 Budapest

• **INDIA**
EWP Affiliated East-West Press
PVT, Ltd
G-I/16, Ansari Road, Darya Gany
New Delhi 110 002
Oxford Book and Stationery Co.
Scindia House, New Delhi 110 001;
17 Park Street, Calcutta 700 016
Oxford Subscription Agency
Institute for Development
Education
1 Anasuya Ave., Kilpauk
Madras 600 010
Periodical Expert Book Agency
D-42, Vivek Vihar, Delhi 110095

• **IRAN**
The FAO Bureau, International and
Regional Specialized
Organizations Affairs
Ministry of Agriculture of the Islamic
Republic of Iran
Keshavarz Bld, M.O.A., 17th floor
Teheran

• **IRELAND**
Publications Section
Government Stationery Office
4-5 Harcourt Road
Dublin 2

• **ISRAEL**
R.O.Y. International
P.O. Box 13056
Tel Aviv 61130

• **ITALY**
Libreria Scientifica Dott. Lucio de
Biasio "Aeiou"
Via Coronelli 6
20146 Milano
Libreria Concessionaria Sansoni
S.p.A. "Licosa"
Via Duca di Calabria 1/1
50125 Firenze

FAO Bookshop
Viale delle Terme di Caracalla
00100 Roma
Tel.   52255688
Fax   52255155
E-mail: publications-sales@fao.org

• **JAPAN**
Far Eastern Booksellers
(Kyokuto Shoten Ltd)
12 Kanda-Jimbocho 2 chome
Chiyoda-ku - P.O. Box 72
Tokyo 101-91
Maruzen Company Ltd
P.O. Box 5050
Tokyo International 100-31

• **KENYA**
Text Book Centre Ltd
Kijabe Street, P.O. Box 47540
Nairobi

• **LUXEMBOURG**
M.J. De Lannoy
202, avenue du Roi
1060 Bruxelles (Belgique)

• **MALAYSIA**
Electronic products only:
Southbound
Sendirian Berhad Publishers
9 College Square
01250 Penang

• **MALI**
Librairie Traore
Rue Soundiata Keita X 115
B.P. 3243
Bamako

• **MAROC**
La Librairie Internationale
70 Rue T'ssoule
B.P. 302 (RP)
Rabat
Tel. (07) 75-86-61

• **MEXICO**
Librería, Universidad Autónoma
de Chapingo
56230 Chapingo
Libros y Editoriales S.A.
Av. Progreso N° 202-1° Piso A
Apdo. Postal 18922, Col. Escandón
11800 México D.F.

• **NETHERLANDS**
Roodveldt Import b.v.
Brouwersgracht 288
1013 HG Amsterdam

• **NEW ZEALAND**
Legislation Services
P.O. Box 12418
Thorndon, Wellington

• **NICARAGUA**
Librería HISPAMER
Costado Este Univ. Centroamericana
Apdo. Postal A-221
Managua

• **NIGERIA**
University Bookshop (Nigeria) Ltd
University of Ibadan
Ibadan

• **NORWAY**
Narvesen Info Center
Bertrand Narvesens vei 2
P.O. Box 6125, Etterstad
0602 Oslo 6
Tel. (+47) 22-57-33-00
Fax (+47) 22-68-19-01

• **PAKISTAN**
Mirza Book Agency
65 Shahrah-e-Quaid-e-Azam
P.O. Box 729, Lahore 3

• **PARAGUAY**
Librería INTERCONTINENTAL
Editora e Impresora S.R.L.
Caballero 270 c/Mcal Estigarribia
Asunción

• **PERU**
INDEAR
Jirón Apurimac 375, Casilla 4937
Lima 1

• **PHILIPPINES**
International Booksource Center (Phils)
Room 1703, Cityland 10
Condominium Cor. Ayala Avenue &
H.V. de la Costa Extension
Makati, Metro Manila

• **POLAND**
Ars Polona
Krakowskie Przedmiescie 7
00-950 Warsaw

• **PORTUGAL**
Livraria Portugal,
Dias e Andrade Ltda.
Rua do Carmo 70-74, Apartado 2681
1117 Lisboa Codex

• **SINGAPORE**
Select Books Pte Ltd
03-15 Tanglin Shopping Centre
19 Tanglin Road
Singapore 1024

• **SOMALIA**
"Samater's"
P.O. Box 936
Mogadishu

• **SOUTH AFRICA**
David Philip Publishers (Pty) Ltd
P.O. Box 23408
Claremont 7735
South Africa
Tel. Cape Town (021) 64-4136
Fax Cape Town (021) 64-3358

• **SRI LANKA**
M.D. Gunasena & Co. Ltd
217 Olcott Mawatha, P.O. Box 246
Colombo 11

• **SUISSE**
Buchhandlung und Antiquariat
Heinimann & Co.
Kirchgasse 17
8001 Zurich
UN Bookshop
Palais des Nations
CH-1211 Genève 1
Van Diermen Editions Techniques
ADECO
Case Postale 465
CH-1211 Genève 19

• **SURINAME**
Vaco n.v. in Suriname
Domineestraat 26, P.O. Box 1841
Paramaribo

• **SWEDEN**
Books and documents:
C.E. Fritzes
P.O. Box 16356
103 27 Stockholm
Subscriptions:
Vennergren-Williams AB
P.O. Box 30004
104 25 Stockholm

• **THAILAND**
Suksapan Panit
Mansion 9, Rajdamnern Avenue
Bangkok

• **TOGO**
Librairie du Bon Pasteur
B.P. 1164
Lomé

• **TUNISIE**
Société tunisienne de diffusion
5, avenue de Carthage
Tunis

• **TURKEY**
Kultur Yayiniari is - Turk Ltd Sti.
Ataturk Bulvari N° 191, Kat. 21
Ankara
Bookshops in Istanbul and Izmir

• **UNITED KINGDOM**
HMSO Publications Centre
51 Nine Elms Lane
London SW8 5DR
Tel. (071) 873 9090 (orders)
      (071) 873 0011 (inquiries)
Fax (071) 873 8463
and through HMSO Bookshops
Electronic products only:
Microinfo Ltd
P.O. Box 3, Omega Road, Alton
Hampshire GU34 2PG
Tel. (0420) 86848
Fax (0420) 89889

• **URUGUAY**
Librería Agropecuaria S.R.L.
Buenos Aires 335
Casilla 1755
Montevideo C.P. 11000

• **USA**
Publications:
UNIPUB
4611/F Assembly Drive
Lanham MD 20706-4391
Toll-free 800 274-4888
Fax 301-459-0056
Periodicals:
Ebsco Subscription Services
P.O. Box 1431
Birmingham AL 35201-1431
Tel. (205)991-6600
Telex 78-2661
Fax (205)991-1449
The Faxon Company Inc.
15 Southwest Park
Westwood MA 02090
Tel. 6117-329-3350
Telex 95-1980
Cable FW Faxon Wood

• **VENEZUELA**
Tecni-Ciencia Libros S.A.
Torre Phelps-Mezzanina
Plaza Venezuela
Caracas
Tel. 782 8697/781 9945/781 9954
Tamanaco Libros Técnicos S.R.L.
Centro Comercial Ciudad Tamanaco
Nivel C-2
Caracas
Tel. 261 3344/261 3335/959 0016
Tecni-Ciencia Libros, S.A.
Centro Comercial, Shopping Center
Av. Andrés Eloy, Urb. El Prebo
Valencia, Ed. Carabobo
Tel. 222 724
Fudeco, Librería
Avenida Libertador-Este
Ed. Fudeco, Apartado 254
Barquisimeto C.P. 3002, Ed. Lara
Tel. (051) 538 022
Fax (051) 544 394
Télex (051) 513 14 FUDEC VC
Fundación La Era Agricola
Calle 31 Junin Qta
Coromoto 5-49, Apartado 456
Mérida
Librería FAGRO
Universidad Central de Venezuela (UCV)
Maracay

• **ZIMBABWE**
Grassroots Books
100 Jason Moyo Avenue
P.O. Box A 267, Avondale
Harare;
61a Fort Street
Bulawayo

Other countries / Autres pays / Otros paises
Distribution and Sales Section
Publications Division, FAO
Viale delle Terme di Caracalla
00100 Rome, Italy
Tel. (39-6) 52251
Fax (39-6) 52253152
Telex 625852/625853/610181 FAO I
E-mail: publications-sales@fao.org